T0234158

Wireless Networks

Series Editor
Xuemin Sherman Shen, University of Waterloo, Waterloo, ON, Canada

The purpose of Springer's Wireless Networks book series is to establish the state of the art and set the course for future research and development in wireless communication networks. The scope of this series includes not only all aspects of wireless networks (including cellular networks, WiFi, sensor networks, and vehicular networks), but related areas such as cloud computing and big data. The series serves as a central source of references for wireless networks research and development. It aims to publish thorough and cohesive overviews on specific topics in wireless networks, as well as works that are larger in scope than survey articles and that contain more detailed background information. The series also provides coverage of advanced and timely topics worthy of monographs, contributed volumes, textbooks and handbooks.

** Indexing: Wireless Networks is indexed in EBSCO databases and DPLB **

Haomiao Yang • Hongwei Li •
Xuemin Sherman Shen

Secure Automatic Dependent Surveillance-Broadcast Systems

Haomiao Yang
University of Electronic Science and
Technology of China
Chengdu, Sichuan, China

Hongwei Li
University of Electronic Science and
Technology of China
Chengdu, Sichuan, China

Xuemin Sherman Shen
University of Waterloo
Waterloo, ON, Canada

ISSN 2366-1186 ISSN 2366-1445 (electronic)
Wireless Networks
ISBN 978-3-031-07023-5 ISBN 978-3-031-07021-1 (eBook)
https://doi.org/10.1007/978-3-031-07021-1

This Springer imprint is published by the registered company Springer Nature Switzerland AG
The registered company address is: Gewerbestrasse 11, 6330 Cham, Switzerland

Preface

The existing air transportation system could not cope well with the expected growth of air traffic in the coming decades. Therefore, the International Civil Aviation Organization (ICAO) launched a thorough system upgrade called the Next Generation Air Transportation System (NextGen). The Automatic Dependent Surveillance-Broadcast (ADS-B) technology is one pillar of NextGen. Surprisingly, the critical ADS-B technology was not designed with sufficient security considerations in mind, making the ADS-B system susceptible to a variety of cyber-attacks because ADS-B messages are publicly transmitted on the wireless broadcast channel and there are no mandatory requirements for encryption and authentication of ADS-B messages. As a result, the security issues of the ADS-B system will raise concerns about the safety of air travel.

To solve ADS-B security problems, in this book, we first conduct a comprehensive vulnerability analysis of ADS-B. Then, we study the security protection approaches of ADS-B based on cryptographic technology and weigh their advantages and disadvantages. On this basis, we propose efficient and compatible ADS-B broadcast authentication, message privacy protection, and aircraft location verification schemes, which fully ensure the security of the ADS-B system and are suitable for large-scale practical deployment. Specifically, our contributions are summarized as follows:

First, we design two ADS-B broadcast authentication schemes, AuthBatch and AuthMR, both of which are based on the identity-based signature. In this case, the public key certificate is no longer needed, thereby significantly saving computational and communication costs for the verification and transmission of the certificate chain. In particular, AuthBatch supports the batch verification of multiple messages, and thus the average verification time for each message is shortened, which can be used in avionics with limited resources. In AuthMR, the message is recovered directly from the signature, so the total length of the message and the signature is reduced, which is suitable for low-bandwidth ADS-B data links. Second, we propose an aircraft location verification scheme that preserves the location privacy of aircraft by using the grid-based k-nearest neighbor algorithm. Specifically, we develop an efficient technique to find k nearest grid squares over ciphertexts based on the vector homomorphic encryption. Further, we design a quick identification method for aircraft legitimacy through validating claimed locations in a small training circle rather than calculating real locations on the entire grid plane, thus greatly reducing the verification time of location claims. Third, although some cryptographic methods have been proposed to protect ADS-B, these methods can only guarantee either privacy or integrity alone, and also need to modify existing ADS-B protocols. To handle these issues, we provide a complete ADS-B security solution to achieve the privacy and integrity of ADS-B messages simultaneously. Furthermore, this solution is not only highly compatible with the existing ADS-B protocol but also appropriate for resource-constrained avionics and congested data links. Besides, it can tolerate packet loss and disorder that are common in ADS-B communications. Extensive tests using real-world flight data show that this solution is deployable in the practical ADS-B system. Finally, considering that cryptographic enhancements may face regulatory and technical complexity, we also suggest several future research directions that will use non-cryptographic techniques to ensure ADS-B security.

We would like to thank Prof. Jianbing Ni, Dr. Dongxiao Liu, Dr. Liang Xue, and Dr. Cheng Huang from the Broadband Communications Research (BBCR) Group at the University of Waterloo; Prof. Xiaosong Zhang, Prof. Yingchang Liang, Prof. Fagen Li, Mr. Mengyu Ge, and Mr. Jiasheng Li from the University of Electronic Science and Technology of China; and Dr. Honggang Wu, Dr. Yi Zhang, and Dr. Zili Xu from the Second Research Institute at the Civil Aviation Administration of China for their contributions to the presented research work. We also would like to thank all the members of the BBCR group for the valuable discussions and their insightful suggestions, ideas, and comments. Special thanks are also due to Mary E. James, Senior Editor at Springer Science+Business Media, for her help throughout the publication process.

This work is also supported by the Key-Area Research and Development Program of Guangdong Province (2020B0101360001), the National Key R&D Program of China (2021YFB3101300, 2021YFB3101302), the National Natural Science Foundation of China (62072081, U2033212), the Fundamental Research Funds for Chinese Central Universities (ZYGX2020ZB027), the CCF-Ant Group Research Fun, and the Sichuan Science and Technology Program (2020JDTD0007).

Chengdu, Sichuan, China   Haomiao Yang
Chengdu, Sichuan, China   Hongwei Li
Waterloo, ON, Canada   Xuemin Sherman Shen

Contents

Acronyms

1090ES	1090 MHz Extended Squitter
AADS	Aviation Asset Distribution System
ADS-B	Automatic Dependent Surveillance-Broadcast
AES	Advanced Encryption Standard
ALV	Aircraft Location Verification
AOA	Angle-of-Arrival
ARTCC	Air Route Traffic Control Center
ATC	Air Traffic Control
ATCO	Air Traffic Controller
CA	Certificate Authority
CBC	Cipher Block Chaining
CFB	Cipher Feedback
CMS	Certificate Management System
CRC	Cyclic Redundancy Check
CRL	Certificate Revocation List
CTR	Counter Mode
DES	Digital Encryption Standard
DoS	Denial of Service
ECB	Electronic Code Book
FAA	Federal Aviation Administration
FIS-B	Flight Information Service-Broadcast
FPE	Format Preserving Encryption
GEO	Geosynchronous Earth Orbit
GNSS	Global Navigation Satellite System
GPS	Global Positioning System
HE	Homomorphic Encryption
IATA	International Air Transport Association
IBC	Identity-Based Cryptography
IBE	Identity-Based Encryption
IBS	Identity-Based Signature
IBS-MR	Identity-Based Signature with Message Recovery

ICAO	International Civil Aviation Organization
IETF	Internet Engineering Task Force
IGSO	Inclined Geosynchronous Orbit
IND-CPA	Indistinguishability under Chosen Plaintext Attacks
ITU	International Telecommunication Union
LSTM	Long Short-Term Memory
MAC	Message Authentication Code
MEO	Medium Earth Orbit
MLAT	Multilateration
NextGen	Next Generation Air Transportation System
NIST	National Institute of Standards and Technology
OFB	Output Feedback
PKG	Private Key Generator
PKI	Public Key Infrastructure
PPM	Pulse Position Modulation
PPT	Probabilistic Polynomial Time
PSR	Primary Surveillance Radar
RA	Registration Authority
SIMT	Single Instruction Multiple Threads
SSR	Secondary Surveillance Radar
TDOA	Time Difference of Arrival
TESLA	Timed Efficient Stream Loss Tolerant Authentication
TIS-B	Traffic Information Service-Broadcast
UAT	Universal Access Transceiver
VAE	Variational Autoencoder
VANET	Vehicular Ad Hoc Network
VHE	Vector Homomorphic Encryption

List of Figures

List of Tables

Chapter 1
Introduction

The Automatic Dependent Surveillance-Broadcast (ADS-B) is progressively displacing radar as the cornerstone technology of the Next Generation Air Transportation System (NextGen) because it may improve air traffic safety by requiring aircraft to broadcast their precise geographic locations frequently [1, 2]. Unfortunately, the lack of security mechanisms, such as message authentication and encryption, significantly hinders the widespread use of this innovative technology [3]. This monograph focuses on how to provide ADS-B with security protection. Specifically, in this chapter, we first outline ADS-B, then introduce two ADS-B data links, the 1090 MHz extended squitter (1090ES) and the universal access transceiver (UAT), and describe the ADS-B message format. Next, we evaluate the vulnerabilities and further define the security needs for ADS-B. Finally, we explain the purpose of the monograph.

1.1 Overview of ADS-B

Nowadays, the demand for air travel is booming. The International Air Transport Association (IATA) predicted that air traffic will double by 2037, reaching 8.2 billion passengers a year.[1] Such much traffic will impose higher traffic density and surveillance complexity on air traffic control (ATC), making air traffic congestion more serious.

To meet the increasing volume for air traffic, the Federal Aviation Administration (FAA) launched the NextGen plan to improve the efficiency and safety of ATC [4]. In particular, NextGen can increase pilots' situational awareness by providing real-time air traffic information and enhancing air traffic conflict identification and resolution. In addition, because the positioning information provided by NextGen is

---

[1] https://www.iata.org/en/pressroom/pr/2018-10-24-02/.

more accurate than that provided by traditional radars, ATC can reduce separations between aircraft. As the core of NextGen, ADS-B contributes most of these improvements [5].

ADS-B requires aircraft to broadcast their location and speed gained through satellite navigation [1]. With the rapid advance of aviation modernization, ADS-B has quickly replaced outdated radar. Because of the ground deployment of radars only detecting within a limited range, quite a few aircraft disappeared at sea (e.g., Malaysia Airlines Flight 370 [6]), which exceeds radar coverage. Unlike traditional radars, emerging ADS-B surveillance gains precise geographic coordinates through the Global Navigation Satellite System (GNSS), which enhances situational awareness and surveillance range, considerably improving flight safety. Since 2020, ADS-B has been compulsory for the airspace of Europe and the USA [7, 8].

1.1.1 Air Traffic Control

Air traffic controllers (ATCOs) on the ground provide ATC services for aircraft surveillance and navigation. They guide aircraft through controlled airspace and advise aircraft flying in uncontrolled airspace. Their primary goals are to avoid aircraft collisions and ensure safe and orderly air traffic. To prevent collisions, ATC makes traffic rules of separation to guarantee each aircraft always preserves the smallest space around it. To this end, ATCOs need to know the identification number, distance, direction, and altitude of aircraft. ATCOs traditionally obtain them through radars, including primary surveillance radars (PSRs) [9] and secondary surveillance radars (SSRs) [10]. PSR works without aircraft cooperation. It sends high-frequency signals reflected by an aircraft, and the reflected echo determines the direction, range, speed, size, and shape of the aircraft. SSR relies on aircraft cooperation, which involves interrogation and response between ground stations and aircraft. The response mainly contains the identification code and location of the aircraft. Today, ADS-B has gradually replaced PSR and SSR, achieving precise positional information through GNSS, extending airspace capacity and increasing air traffic safety. Specifically, ADS-B provides the following enhancements:

- Reducing the risk of aircraft crashes
- Reducing the working cost of ATC
- Improving the situational awareness of pilots
- Improving the capacity to handle drones
- Improving positioning and speed accuracy to within 100 m and 10 m a second, respectively
- Improving worldwide coverage containing transoceanic regions [11]

Such enhancements also reduce traffic congestion by allowing more aircraft to fly in the same airspace and low flight delay by using more direct routes. Therefore, the running cost of ADS-B is significantly lower than radar. For monitoring airspace with a 200-nautical-mile radius, PSR, Mode S, and ADS-B need to take $10–

Table 1.1 Comparison of SSR and ADS-B

	SSR	ADS-B
Positioning	Time of arrival	GNSS
Operating cost	High	Low
Global coverage	✗	✓
Capacity improvement	✗	✓
Aircraft-to-aircraft	✗	✓
Aircraft intention	✗	✓
Separation guarantee	✗	✓

14,000,000, $6,000,000, and $380,000, respectively [12]. Table 1.1 briefly compares ADS-B and SSR, where the symbols of "✓" and "✗" show if there is functionality.

1.1.2 ADS-B Characteristics

The name of ADS-B (Automatic Dependent Surveillance-Broadcast) reflects its basic characteristics: ADS-B automatically works without operator intervention, depends on GNSS to extract precise location information, provides radar-like surveillance services, and periodically broadcasts the status of aircraft. Especially, the ADS-B system has the following characteristics [13]:

– The ADS-B communication mainly comprises unidirectional broadcasts. Current implementations of ADS-B communications still use single-hop unidirectional broadcast links, although research on multi-hop communications is increasing.
– Due to the compulsory implementation of ADS-B, the channel utilization of ADS-B has increased rapidly, leading to channel congestion. Therefore, packet loss becomes a major problem, especially in high-density airspace. However, the ADS-B protocol does not deal with packet loss but hands this problem to the high-level protocol for processing.
– The ADS-B network is dynamic with high mobility. Aircraft usually fly at a speed of 700–1000 km an hour, and communication between two mobile aircraft via ADS-B may only last a few seconds. Besides, since the trajectory of the aircraft is not subject to physical limits, the ADS-B network is still effective at distances of 100 nautical miles and above.

1.1.3 ADS-B System Framework

As Fig. 1.1 shows, ADS-B includes two communication subsystems, ADS-B-Out and ADS-B-In. Using an airborne transmitter, ADS-B-Out provides more accurate positions than radar, and then ATCO can separate aircraft within a shorter distance,

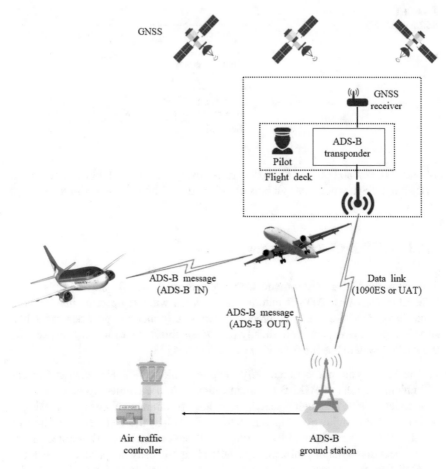

Fig. 1.1 ADS-B system framework

increasing airspace capacity. ADS-B-In allows aircraft to receive nearby ADS-B messages comprising the aircraft's latitude, longitude, altitude, speed, transponder code, call sign, weather information, airspace status information, etc. This informa-tion is displayed in the cockpit to raise the pilot's situational awareness.

Running ADS-B relies on two avionics components: the GNSS for positioning and the datalink for communication. In particular, airborne ADS-B avionics include the GNSS receiver, ADS-B transceiver, antenna, multifunction cockpit, etc.

GNSS refers to a satellite constellation offering navigation, positioning, and timing services on a global or regional scale [14]. Although the Global Positioning System (GPS) is the most commonly used GNSS, other countries have deployed or are deploying their GNSS such as BeiDou to provide complementary independent positioning capabilities. Table 1.2 compares three common GNSS systems: GPS, GLONASS, and BeiDou.

Table 1.2 Comparison of GPS, GLONASS, and BeiDou

	GPS	GLONASS	BeiDou
Number of satellites	31	24	14 (5 GEO, 5 IGSO, 4 MEO)
Number of nominal satellites	24	24	35 (5 GEO, 3 IGSO, 27 MEO)
Number of orbital planes	6	3	3 (MEO)
Inclination plane	55°	65°	55° (MEO & IGSO)
Altitude (km)	20,200	19,140	21,530 (MEO)
Orbital period	11 h 58 min	11 h 16 min	12 h 50 min (MEO)
Time scale	GPST UTC (USNO)	UTC (SU)	BDT UTC (NTSC)
Coordinate system	WGS 84	PZ 90	CGCS 2000
Ephemerides	Kepler Elements	Geocentric Cartesian	Kepler Elements
Ephemeris update	Every 2 h	Every 0.5 h	Every 1 h
Message length	12.5 min	2.5 min	12 min

- *US (GPS).* GPS comprises 24 satellites, each of which orbits the Earth at 20,200 km. For tracking, two GPS satellites are outfitted with laser retroreflectors. The deployment of 24 satellites allows six of them to be observed almost 100% of the time from anywhere on the planet.
- *Russia (GLONASS).* For satellite constellation, orbit, and signal, GLONASS is like GPS, also including 24 satellites, each of which orbits 19,140 km from the earth. These satellites are all outfitted with laser retroreflectors, easily performing ranging with satellites.
- *China (BeiDou).* BeiDou includes 35 satellites covering geosynchronous earth orbit (GEO), inclined geosynchronous orbit (IGSO) and medium earth orbit (MEO). BeiDou works in China and serves customers in the Asia-Pacific region.
- *European Union (Galileo).* When completely deployed, Galileo will include 27 operational and 3 spare satellites, operating at a height of 23,222 km in three orbital planes. These satellites will be all outfitted with laser retroreflectors.
- *Japan (QZSS).* QZSS will deploy 3 satellites in multiple orbital planes for regional positioning and time transfer. It provides a high-precision satellite positioning service covering for almost all areas of Japan, including cities and mountainous areas.

For ADS-B data links, there are two competing standards: UAT and 1090ES [15]. As shown in Table 1.3, UAT is designed for a variety of aviation services, including ADS-B. Because of the need to install new hardware, UAT is currently only used for general aviation in US airspace. Commercial aircraft mainly adopt the

Table 1.3 Comparison of ADS-B data links

	UAT	1090ES
Scope	US airspace	Worldwide airspace
Standard	ICAO DOC-9861	RTCA DO-260C
	RTCA DO-282B	EUROCAE ED-129B
Motivation	Overcoming the limitation of 1090ES	Improving SSR
Service	FIS-B/TIS-B/ADS-B	ADS-B
Frequency	978 MHz	1090 MHz
Transfer rate	1.04 Mbps	1 Mbps
Bandwidth	1.3 MHz	11 MHz
Pros	Two-way service support	High channel error correction capability
	Resistance to fading	High Compatibility with Mode S transponder
Cons	Requiring new devices	No support for two-way services
	Requiring localized frequencies	Limited potential for growth

1090ES, which integrates ADS-B into the traditional Mode S transponder and is recommended by ICAO for global adoption. Today, 1090ES has been recognized and applied in most countries, such as the USA, Europe, and Asia. From now on, unless otherwise stated, references to data links in this monograph refer to the 1090ES.

1090ES Data Link The 1090ES has two meanings: 1090 means the data link employs a 1090 MHz frequency of downlink; ES refers to ADS-B messages attached to Mode S data through an extended squitter [16]. The 1090ES is a Mode S data link supporting one-to-one interrogation between aircraft and ground stations. In particular, the S stands for selective interrogation. The 1090ES uses pulse position modulation (PPM) coding to send different types of ADS-B messages, such as ICAO address, position, and speed. Due to the weak ability to carry information, a code can only transmit a specific message. The update rate of different messages is also different. For example, the position and velocity are updated every 400–600 ms, while the identifier is updated every 480–520 ms.

UAT Data Link UAT was launched after 1090ES [17], which works as a single-channel broadband data link at a frequency of 978 MHz and can send data at a rate of about 1 Mbps. The media access of UAT divides a single-second frame into two parts, allowing for multi-access: the first is for ground-based broadcast services (TIS-B and FIS-B), while the second is for ADS-B. Notably, the Traffic Information Service-Broadcast (TIS-B) provides ADS-B-equipped aircraft with surveillance data from non-ADS-B-equipped aircraft. The Flight Information Service-broadcast (FIS-B) is an uncontrolled surface-to-air broadcast service that provides pilots with alert data for safety and efficiency. The FAA uses UAT as an alternative to Mode S for the controlled airspace in the USA. UAT works on a frequency of 978 MHz,

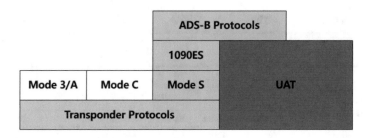

Fig. 1.2 Hierarchy of ADS-B data links

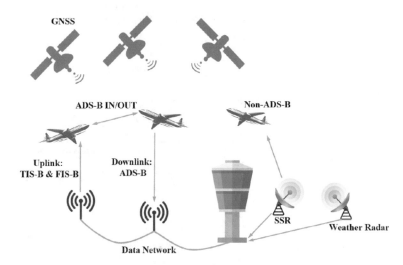

Fig. 1.3 Aircraft communication based on ADS-B

different from the 1090 MHz of Mode S, relieving signal congestion in Class B airspace. Figure 1.2 shows the ADS-B protocol level, where UAT only applies to general aviation, while 1090ES can be widely used in commercial aviation. Figure 1.3 further shows the integrated aircraft communication based on ADS-B.

1.1.4 ADS-B Message Format

Mode S specifies two optional message lengths, 56 bits and 112 bits, while ADS-B only uses the 112-bit long format. Figure 1.4 illustrates the long format of 1090ES, starting from a synchronization preamble and following a data block. In the data block, the downlink format (DF) field sets message types. The 1090ES adopts a

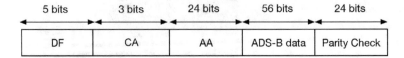

5 bits	3 bits	24 bits	56 bits	24 bits
DF	CA	AA	ADS-B data	Parity Check

Fig. 1.4 1090ES message format

Table 1.4 Field meaning of 1090ES

Field	Value	Meaning
DF	17	Mode S transponder
	18	Non-Mode S transponder
	19	Military purpose
CA	DF = 17	Capabilities of Mode S transponder
AA		ICAO address
DATA		ADS-B data
PC		Parity check

Table 1.5 Type codes of 1090ES

Type code	Meaning
1–4	Aircraft identification
5–8	Surface position
9–18	Airborne position (w/Baro Altitude)
19	Airborne velocity
20–22	Airborne position (w/GNSS Height)
23	Test
24	Surface system status
25–27	Reserved
28	Extended squitter AC status
29	Target state and status (V.2)
30	Reserved
31	Aircraft operation status

multi-purpose format, especially when DF = 17, the message is an extended squitter, allowing 56-bit application data to be filled in the ADS-B data (DATA) field. The Capability (CA) field represents the transponder ability, and the 24-bit aircraft address (AA) field carries a unique ICAO address identifying the aircraft. Finally, the parity check (PC) field provides a 24-bit cyclic redundancy check (CRC) to detect and correct transmission errors. Specifically, a $24°$ generator polynomial is constructed to correct errors of up to 5 bits [18]. Table 1.4 examines the meaning of all fields.

Besides, the first 5 bits of the 56-bit DATA field are type codes defining the types of ADS-B messages. Table 1.5 shows the meaning of the specific type code.

1.1.5 ADS-B Protocol Stack

Figures 1.5 and 1.6 illustrate ADS-B protocol stacks of air-to-air and air-to-ground communications, respectively. We divide these protocols into two categories: aircraft-based and ground station-based. The former includes three layers: airborne application, ADS-B-Out/In, and airborne radio. The latter accordingly comprises three layers: ground application, ADS-B-In/Out, and ground radio. Further, the radio layer of either the former or the latter consists of 3 sub-layers: message assembly, frame assembly, and radio frequency (RF). On the broadcast side, the radio layer first gets application data, then assembles the frame containing the ADS-B message, and finally modulates and broadcasts ADS-B signals. At the receiving end, the radio layer demodulates received signals, extracts ADS-B messages, and submits them to the application layer for further processing.

Fig. 1.5 ADS-B air-to-air protocol stack

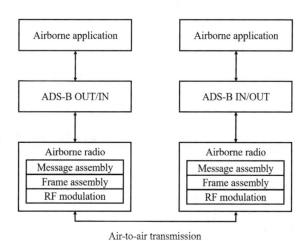

Air-to-air transmission

Fig. 1.6 ADS-B air-to-ground protocol stack

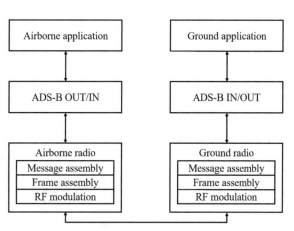

1.2   ADS-B Vulnerabilities

Because of no security protection, ADS-B is susceptible to various cyber-attacks. For instance, the mobile application called *Plane Finder AR* may in real-time offer any flight information; at the 2012 Black-Hat conference, Costin and Francillon [19] proved we can easily insert a fake aircraft into the surveillance screen using a cheap ADS-B transmitter. The ICAO organization has learned of ADS-B vulnerabilities and recommended ADS-B security research to discover possible cryptographic solutions. Besides, because ADS-B is built on GNSS, it also inherits GNSS vulnerabilities. In this section, we will discuss ADS-B vulnerabilities from passive attacks, active attacks, and GNSS flaws [20–22].

1.2.1   Passive Attacks

The primary form of passive attacks in ADS-B is eavesdropping, also called *Aircraft Reconnaissance*, which is an act of listening to ADS-B broadcasts [23]. Since the ADS-B-Out communication subsystem broadcasts in the clear, anyone with an ADS-B receiver can get ADS-B messages. Eavesdropping on ADS-B broadcasts can intercept important ADS-B messages such as locations of *Air Force One* in flight. Terrorists can then track the presidential plane and launch attacks. Other companies can also link the trajectory of a competitor's jet to its business strategy. Moreover, eavesdropping is the basis of more sophisticated active attacks [24]. Unfortunately, eavesdropping is impossible to detect, so some countries have to establish regulations to avoid eavesdropping on public broadcast traffic. It is also difficult to prevent eavesdropping without ADS-B message encryption, but the encryption is inconsistent with the openness of the ADS-B system. For example, the FAA claims the need for plaintext ADS-B messages for flight safety.

1.2.2   Active Attacks

Active attack to ADS-B is a way to create chaos and terror in the air. An adversary in an active attack tries to affect flight operations, which involves changing ADS-B messages or creating false statements. The types of active attacks are as follows.

Jamming Jamming can flood ADS-B receivers by emitting high enough power on the 1090 MHz frequency of Mode S. Jamming may interfere with a specific aircraft, a specific ground station, and even the entire area covering ground stations and aircraft. By doing so, jamming can effectively disable all ADS-B message traffic in the affected area [25]. Although interference is a common problem in wireless communications, the impact of interference on aircraft communications is even more serious due to airspace complexity and communication importance. Jamming a single ground station is a low-difficulty attack. Schafer et al. [26] demonstrated this attack by sending continuous white noise to a ground station, resulting in a denial of service at the ground station. It would be difficult to black out an area distributing various ADS-B receivers due to high-power interference. A powerful jamming attack could devastate a high-density area if the jamming causes collision avoidance to fail. It is more difficult to jam a moving aircraft because of the aircraft's high-speed mobility.

Message Injection Due to the lack of authentication for ADS-B messages, an attacker may inject illegal ADS-B messages into the ADS-B system without being detected. For example, this attack can create ghost planes that are visible to both pilots and ATCO, making pilots interact with these fake planes, or ATCO unnecessarily guide the fake planes or refuse to let the real planes land. Fake messages are hard to distinguish from real ones since they also have the correct format, and thus any ADS-B receiver will treat these false messages as legitimate ones. Here, a single ghost airplane is difficult to separate from a real one, so myriad ghost airplanes would flood the skies, resulting in a complete denial of ATC service. Schafer et al. proved the feasibility of ghost airplane flooding using a software-defined radio connected to a computer flight simulator [26].

Message Deletion This deletion attack uses constructive or destructive interference to make ADS-B messages disappear from the radio frequency medium. Destructive interference synchronously mixes inverse signals into original ADS-B signals to superimpose, thereby canceling or significantly attenuating legitimate signals. This attack requires precise and complex timing, making it extremely challenging in practice. Constructive interference requires no precise timing synchronization and causes errors in transmitted messages. Since 1090ES can correct up to 5-bit errors for each message, messages beyond this limit will be automatically discarded by ADS-B receivers as corrupted, effectively deleting the messages. This attack needs careful positioning of the interference transmitter to successfully carry out message deletion far from 100 km [3]. Further, the attacker can change an aircraft's position by combining message deletion with injection. The attacker first deletes all messages from the aircraft and then injects messages seemingly from the aircraft with the changed position. For a successful attack, injected messages have higher transmission power than deleted messages. Besides location modification, this attack may also inject messages, showing the existing aircraft is in an emergency or hijacked. The false emergency or hijacking will seriously mislead ATC [26].

Message Modification Message modification on the physical layer comprises two ways: overshadowing and bit-flipping. Overshadowing uses high-power signals to replace all or part of targeted messages. Bit-flipping superimposes a signal on the original signal and converts multiple bits from 0 to 1 and vice versa. In both cases, arbitrary data can be injected with no participant's knowledge. This effect can also be achieved by combining message deletion with injection. However, since manipulated messages are initially legitimate, message modification may be more dangerous than injecting brand-new messages. The works in [27] and [28] have proved the feasibility for such message modification.

1.2.3 GNSS Vulnerabilities

The reliance of ADS-B on GNSS ensures the accuracy of ADS-B broadcast information. However, the open publication of GNSS frequencies and the clear-text civilian signal of GNSS broadcast result in GNSS vulnerabilities, which worsen air traffic safety. GNSS has two significant vulnerabilities: jamming [29] and spoofing [30]. At the same time, the normal GPS signal is weak, only between 10 and 16 watts, roughly equating to the light from a 20-watt light bulb when viewed from 12,000 miles away.

GNSS Jamming GPS may be accidentally interfered with by solar flares or other radio frequency signals with frequencies close to GPS signals. GPS may also be deliberately interfered with by GPS jammers, and the jamming signals will overwhelm the signals generated by GPS satellites. For example, when a car lighter is plugged in, a small jammer will interfere with GPS signals a quarter of a mile away, and a jammer with the size of a soda can will interfere with GPS signals 40 miles away [31]. Here, ADS-B receivers can no longer get locations from GPS [32].

GNSS Spoofing If the perpetrator uses a GPS simulator, spoofing the GPS may be a simple process. GPS simulators produce the same signals as GPS satellites, and GPS receiver manufacturers use them to test the accuracy of their products. Since their signals are the same as those produced by GPS satellites, they can change the position calculated by the GPS receiver. In particular, GPS simulators use the same identification numbers broadcast by satellites to simulate real GPS signals. The GPS simulator operator slowly increases the power transmitted by the GPS simulator until the target GPS receiver locks onto a stronger signal and ignores the original signal. Using higher power and manipulating the transmission time of fraudulent signals, a spoofing attack will slowly change the position calculated by the GPS receiver, causing the operator of the target aircraft to deviate from its planned route.

1.2.4 Security Risk Assessment

Attackers may cause havoc with ATC. To conduct a more thorough analysis of the risks associated with assaults, first it is helpful to identify potential ADS-B attackers. There are three kinds of attackers:

- *Insider attacker.* These are individuals with malevolent intent who work at an airport, a ground station, or aboard an aircraft. Such employees have physical access to restricted areas and information, as well as related equipment and

Table 1.6 Attack classification according to security goals

	Confidentiality	Integrity	Authentication	Availability
Eavesdropping	✓			
Jamming				✓
Message injection		✓	✓	
Message modification		✓		
Message deletion		✓		

networks. Therefore, the possibility of attacks, such as message modification, deletion, or interference, will increase. In particular, if the attack targets aircraft, there is a greater chance of success for an insider attacker than for a remote attacker.

– *Malicious passenger.* They can launch ADS-B attacks on the aircraft during the flight. The risk of a successful attack caused by them is less than that of insiders.
– *Remote attacker.* These attackers can launch attacks from unknown locations, not at airports or airplanes. They can effectively perform denial of service (DoS) attacks and endanger the ATC in their coverage area. When the target aircraft exceeds the coverage range of an attacker, the attacks are no longer effective, so it is less likely to attack a specific aircraft during the entire flight. The possibility of a remote attacker is obviously higher than the other two kinds.

Table 1.6 classifies ADS-B vulnerabilities based on information security goals: confidentiality, integrity, authentication, and availability.

– *Confidentiality.* Only authorized parties can access sensitive aviation data. Any act causing information disclosure to unauthorized parties violates confidentiality, so this class includes ADS-B eavesdropping attacks.
– *Integrity.* During transmission, ADS-B messages must not be changed, so we classify message deletion and modification as attacks on integrity.
– *Authentication.* During ADS-B communications, the recipient should be able to authenticate the identity of the ADS-B message sender. By inserting messages into actual communications, the attacker may broadcast ADS-B messages with unauthenticated identities.
– *Availability.* Allowed parties should be able to access ADS-B messages and services. Unfortunately, an attacker can easily interfere with ADS-B ground stations and launch DoS attacks. We thus include jamming in this class.

Table 1.7 classifies attacks on ADS-B messages by considering the compromised ADS-B protocol layer. The message assembly sub-layer forms ADS-B messages bit-by-bit, so message modification occurs at this sub-layer. Since message injection (or deletion) adds (or deletes) messages to (or from) the transmitted frame, they destroy the frame assembly sub-layer. Eavesdropping and jamming are carried out in RF modulation, and thus we classify them in the RF modulation sub-layer.

We further analyze ADS-B vulnerabilities based on the possibility of attack versus the severity of the attack result. Figure 1.7 summarizes the risk assessment

Table 1.7 Attack classification according to protocol layers

	Message assembly	Frame assembly	RF transmission
Eavesdropping			✓
Jamming			✓
Message injection		✓	
Message modification	✓		
Message deletion		✓	

Fig. 1.7 Risk assessment matrix illustrating attack possibility versus impact

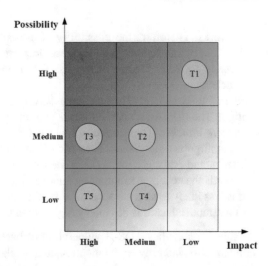

matrix concerning attacks. We use the symbols T1, T2, T3, T4, and T5 to represent eavesdropping, jamming, message injection, message deletion, and message modification, respectively. Specifically, we have the following observations:

- *T1*. Due to no message encryption, it is easy to eavesdrop on transmitted messages, but since T1 has no effect on ATC, its risk is slight.
- *T2*. This attack is more likely to occur because the attacker can easily approach the ground station and interfere with the ADS-B channel in the area. Since the impact of T2 is limited to a single area and is proportional to the strength of the interfering signal, we consider it to have a moderate impact.
- *T3*. If this attack is to distract pilots and ATCOs, the likelihood is moderate, but the impact is small. If attackers inject many fake aircraft, T3 may have a significant impact because it will disrupt air traffic, mislead collision avoidance, and cause accidents. In this case, we consider T3 to be a major danger to ATC.
- *T4*. This attack requires time synchronization, so it is difficult to carry out, which reduces the attack likelihood. The impact of this attack on ATC is also limited because, even if this attack removes aircraft from the ATC screen, ATC can still be supported by a backup system (such as multilateration), reducing the severity of the attack.

- *T5*. T5 may hijack aircraft remotely and cause an accident, so it has a significant impact on ATC. But the likelihood of this attack is slight because of high complexity and strict time synchronization.

Finally, ADS-B attackers may change ADS-B messages, delete real messages, inject deceptive messages, or interfere with communication channels to reduce communication reliability between aircraft and ATCO. Untrusted data provided to other information systems has a negative impact on overall flight management. Therefore, the hacked system has a domino effect on more aviation systems. Below, we list some of the affected systems: [33]:

- Systems of communication, surveillance, and navigation
- Systems of air traffic aided management
- Systems of flight tracking, monitoring, and management
- Systems of departure control
- Systems of airline-passenger communication

1.3   ADS-B Security Requirements

Concerning ADS-B vulnerabilities, an ADS-B security solution should meet the following requirements:

- *Privacy.* An unauthorized third party should not associate an aircraft's ADS-B message with its digital identity nor should it connect the aircraft's two consecutive messages.
- *Integrity.* The received message should be the same as the provided one, not altered by a third party. Also, the message originates from the participant who claims to have sent it. Besides, the message comes from the location specified in the message.
- *Compatibility.* This solution should be compatible with existing ADS-B installations and should not modify hardware or software standards.
- *Scalability.* This solution should be easy to extend to increase local aircraft density and global aircraft traffic.
- *Tolerance to packet loss.* This solution should be able to tolerate packet loss even for crowded wireless channels.

1.4   Aim and Organization of Monograph

Vulnerability analysis of the ADS-B system can help readers understand the full scope of ADS-B security issues, so aviation experts, computer security analysts, and government officials can better realize these vulnerabilities and support them in proposing ideas for feasible ADS-B security solutions. It is also beneficial to expose the weaknesses of NextGen and set off a wave of advocacy for the greater protection of the ADS-B system. In particular, we propose some promising ADS-B security solutions based on cryptography, which can be used as a reference for air traffic safety decision-making in the industry and related government departments. This monograph is organized as follows:

- In Chap. 1, we outline the ADS-B system, followed by an introduction to ADS-B data links containing 1090ES and UAT. We then assess ADS-B vulnerabilities and accordingly specify the security requirements for the ADS-B system.
- In Chap. 2, we briefly introduce modern cryptography and then evaluate the usability of cryptography in the ADS-B system. Finally, according to the security requirements of the ADS-B system, we discuss some emerging cryptographic primitives suitable for ADS-B security.
- In Chap. 3, we propose two ADS-B broadcast authentication schemes, Auth-Batch and AuthMR. The former can be used for avionics with limited resources, while the latter is suitable for low-bandwidth ADS-B data links. Specifically, both are based on IBS and no longer need PKI, thus reducing the costs of key management, and certificate verification and transmission.
- In Chap. 4, we propose a secure and efficient aircraft location verification scheme, ALV, which is based on the grid-based kNN technique, significantly improving the accuracy of ALV. Additionally, we design a quick ALV method, which greatly reduces the time required for verification.
- In Chap. 5, we solve the unique security problems of the realistic ADS-B environment and provide a complete cryptographic solution to ensure the privacy and integrity of ADS-B communication. At the same time, the solution has good compatibility with the current ADS-B protocol. A large number of tests using real-world flight data show that the solution is suitable for practical deployment in practical ADS-B systems.
- In Chap. 6, we conclude this monograph and investigate several non-cryptographic research topics in ADS-B security for future work.

References

1. RTCA, "Minimum aviation system performance standards for automatic dependent surveillance broadcast (ads-b)," October 2002.
2. E. Valovage, "Enhanced ads-b research," in *2006 IEEE/AIAA 25TH Digital Avionics Systems Conference*. IEEE, 2006, pp. 1–7.
3. M. Strohmeier, M. Schäfer, V. Lenders, and I. Martinovic, "Realities and challenges of NextGen air traffic management: the case of ADS-B," *IEEE Communications Magazine*, vol. 52, no. 5, pp. 111–118, 2014.
4. P. Brooker, "Sesar and nextgen: Investing in new paradigms," *The Journal of Navigation*, vol. 61, no. 2, pp. 195–208, 2008.
5. T. L. Davis, "Global SESAR1/NextGen internet based ATN infrastructure," in *Proceedings of IEEE ICNS 2016*, 2016, pp. 1–17.

6. P. P. Pan and K. Semple, "A routine flight, till both routine and flight vanish," https://www.nytimes.com/2014/03/23/world/asia/a-routine-flight-till-both-routine-and-flight-vanish.html, March 2014.

7. EUROCONTROL, "Cascade news 9 - update on developments," October 2010.

8. FAA, "Automatic dependent surveillance–broadcast," October 2016.

9. N. Young, R. Hayward, and D. Dow, "Multi-static primary surveillance radar: An air navigation service provider perspective," in *2015 16th International Radar Symposium (IRS)*. IEEE, 2015, pp. 266–271.

10. I. Svyd, I. Obod, O. Maltsev, I. Shtykh, G. Zavolodko, and G. Maistrenko, "Model and method for request signals processing of secondary surveillance radar," in *2019 IEEE 15th International Conference on the Experience of Designing and Application of CAD Systems (CADSM)*. IEEE, 2019, pp. 1–4.

11. K. D. Wesson, T. E. Humphreys, and B. L. Evans, "Can cryptography secure next generation air traffic surveillance?" *IEEE Security and Privacy Magazine*, 2014.

12. I. Cao, "Guidance material on comparison of surveillance technologies (GMST)," *International Civil Aviation Organization*, 2007.

13. K. Pourvoyeur, A. Mathias, and R. Heidger, "Investigation of measurement characteristics of MLAT/WAM and ADS-B," in *2011 Tyrrhenian International Workshop on Digital Communications-Enhanced Surveillance of Aircraft and Vehicles*. IEEE, 2011, pp. 203–206.

14. H. Park, A. Camps, J. Castellvi, and J. Muro, "Generic performance simulator of spaceborne GNSS-reflectometer for land applications," *IEEE Journal of Selected Topics in Applied Earth Observations and Remote Sensing*, vol. 13, pp. 3179–3191, 2020.

15. Y. H. Chen, S. Lo, P. Enge, and S. S. Jan, "Evaluation & comparison of ranging using universal access transceiver (UAT) and 1090 MHz Mode S extended squitter (Mode S ES)," in *Proceedings of IEEE/ION PLANS 2014*, 2014, pp. 915–925.

16. W. Harman, J. Gertz, and A. Kaminsky, "Techniques for improved reception of 1090 MHz ads-b signals," in *17th DASC. AIAA/IEEE/SAE. Digital Avionics Systems Conference. Proceedings (Cat. No. 98CH36267)*, vol. 2. IEEE, 1998, pp. G25–1.

17. S. Lo and Y.-H. Chen, "Automatic dependent surveillance-broadcast (ads-b) universal access transceiver (UAT) transmissions for alternative positioning, navigation, and timing (APNT): Concept & practice," *NAVIGATION, Journal of the Institute of Navigation*, vol. 68, no. 2, pp. 293–313, 2021.

18. A. Abdulaziz, A. S. Yaro, A. A. Adam, M. T. Kabir, and H. B. Salau, "Optimum receiver for decoding automatic dependent surveillance broadcast (ads-b) signals," *American Journal of Signal Processing*, vol. 5, no. 2, pp. 23–31, 2015.

19. D. Storm, "Curious hackers inject ghost airplanes into radar, track celebrities' flights," http://www.computerworld.com/article/2472455/cybercrime-hacking/curious-hackers-inject-ghost-airplanes-into-radar.html, August 2012.

20. M. U. Iqbal and S. Lim, "Legal and ethical implications of GPS vulnerabilities," *J. Int'l Com. L. & Tech.*, vol. 3, p. 178, 2008.

21. L. Purton, H. Abbass, and S. Alam, "Identification of ads-b system vulnerabilities and threats," in *Australian Transport Research Forum, Canberra*, 2010, pp. 1–16.

22. M. R. Manesh and N. Kaabouch, "Analysis of vulnerabilities, attacks, countermeasures and overall risk of the automatic dependent surveillance-broadcast (ads-b) system," *International Journal of Critical Infrastructure Protection*, vol. 19, pp. 16–31, 2017.

23. D. McCallie, J. Butts, and R. Mills, "Security analysis of the ads-b implementation in the next generation air transportation system," *International Journal of Critical Infrastructure Protection*, vol. 4, no. 2, pp. 78–87, 2011.

24. M. Schäfer, M. Strohmeier, V. Lenders, I. Martinovic, and M. Wilhelm, "Bringing up OpenSky: A large-scale ads-b sensor network for research," in *IPSN-14 Proceedings of the 13th International Symposium on Information Processing in Sensor Networks*. IEEE, 2014, pp. 83–94.

25. M. Wilhelm, I. Martinovic, J. B. Schmitt, and V. Lenders, "Short paper: reactive jamming in wireless networks: how realistic is the threat?" in *Proceedings of the fourth ACM conference on Wireless network security*, 2011, pp. 47–52.
26. M. Schäfer, V. Lenders, and I. Martinovic, "Experimental analysis of attacks on next generation air traffic communication," in *International Conference on Applied Cryptography and Network Security*. Springer, 2013, pp. 253–271.
27. C. Pöpper, N. O. Tippenhauer, B. Danev, and S. Capkun, "Investigation of signal and message manipulations on the wireless channel," in *European Symposium on Research in Computer Security*. Springer, 2011, pp. 40–59.
28. M. Wilhelm, J. B. Schmitt, and V. Lenders, "Practical message manipulation attacks in IEEE 802.15. 4 wireless networks," in *MMB & DFT 2012 Workshop Proceedings*, 2012, pp. 29–31.
29. M. Leonardi, E. Piracci, and G. Galati, "Ads-b vulnerability to low cost jammers: Risk assessment and possible solutions," in *2014 Tyrrhenian International Workshop on Digital Communications-Enhanced Surveillance of Aircraft and Vehicles (TIWDC/ESAV)*. IEEE, 2014, pp. 41–46.
30. M. L. Psiaki and T. E. Humphreys, "GNSS spoofing and detection," *Proceedings of the IEEE*, vol. 104, no. 6, pp. 1258–1270, 2016.
31. J. Keller, "Gps jamming is a growing threat to satellite navigation, positioning, and precision timing?" *Military & Aerospace Electronics*, 2016.
32. S. Cole, "Securing military GPS from spoofing and jamming vulnerabilities," *Military Embedded Systems*, 2015.
33. S. S. Airports, "European union agency for network and information security, 2010," *DOI*: https://doi.org/10.2824/865081, 2010.

Chapter 2
Modern Cryptography for ADS-B Systems

Modern cryptography has been proven to be a mature technology to protect the security of wireless communication, so it should be workable to secure ADS-B wireless broadcasts. Specifically, by evaluating the practicability and effectiveness of cryptographic methods in the aviation industry, where technology is complex and cost-saving, it can be concluded that cryptography can indeed safeguard the ADS-B system [1, 2]. In this chapter, we first briefly introduce modern cryptography and then discuss some cryptographic primitives that are proper for securing ADS-B.

2.1 Overview of Modern Cryptography

As the backbone of information and communication technology (ICT), modern cryptography has three main characteristics that are different from classic ciphers [3]:

- It uses publicly available mathematical methods to encode data, and secrecy is achieved by using the key as the seed of the algorithm. Because of the computational complexity of such encoding and the lack of keys, even if adversaries are aware of the encoding method, they cannot recover the plaintext data.
- It requires parties who wish to communicate securely to hold the key.
- It deals with binary bit sequences, not traditional characters.

Below, we list some commonly used modern cryptography techniques.

© The Author(s), under exclusive license to Springer Nature Switzerland AG 2023
H. Yang et al., *Secure Automatic Dependent Surveillance-Broadcast Systems*,
Wireless Networks, https://doi.org/10.1007/978-3-031-07021-1_2

2.1.1 Symmetric Ciphers

Unlike traditional characters (letters and numbers), digital messages are represented by binary bit strings, and modern encryption schemes are required to manipulate (encrypt and decrypt) these bit strings to perform the conversion between plaintext and ciphertext. Especially in symmetric ciphers, only one key is used for encryption and decryption. According to the processing methods of binary strings, symmetric ciphers are divided into two categories: block ciphers [4] and stream ciphers [5], as shown in Fig. 2.1.

For stream ciphers, the plaintext is processed bit-by-bit; that is, one bit in the plaintext corresponds to one bit in the ciphertext using the stream cipher. A stream cipher can be technically considered as a block cipher with a block size of one bit. For block ciphers, the plaintext binary text is processed in bit blocks at a time; that is, a plaintext bit block is selected and a series of operations are performed on it to generate a ciphertext bit block. The number of bits in the block is fixed. For example, the block sizes of the Digital Encryption Standard (DES) and the Advanced Encryption Standard (AES) are 64 and 128 bits, respectively [6]. Considering that block ciphers are commonly used in ADS-B environments, here we discuss two important details about block ciphers.

– *Block Size*. Although any block size can be accepted in theory, the following factors will be considered when selecting block sizes. Assuming the size of the block is m bits, the number of potential plaintext bit combinations is 2^m. If an attacker learns a plaintext block that matches the previously transmitted ciphertext block, he/she can perform a dictionary attack by compiling a plaintext/ciphertext paired dictionary. Fortunately, attacking larger blocks requires larger dictionaries, and when the block size exceeds a certain threshold, the attack becomes inefficient.

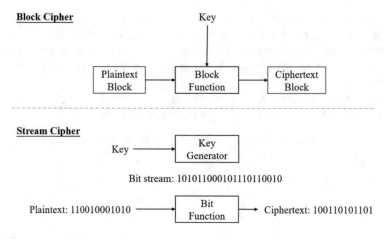

Fig. 2.1 Symmetric ciphers

- *Padding.* The block cipher processes blocks of a specified size (for example, 64 bits). However, the length of the plaintext is usually not a multiple of the block size. For example, a 150-bit plaintext provides two 64-bit blocks and a third 22-bit block. The last bit block must be filled with additional information so that its length corresponds to the block size required by the block cipher. In this case, an additional 42 bits need to be added after the remaining 22 bits to create a complete block. Padding is the process of adding bits to the last block. When running a block cipher algorithm, overfilling the system can lead to inefficiency. In addition, padding can sometimes make the system insecure, especially if padding always uses the same bits.

2.1.2   Advanced Encryption Standard

Commonly used symmetric-key encryption algorithms include DES, 3DES (Triple DES), AES, RC2, RC4, RC5, Blowfish, etc. Among them, the most frequently used is the Advanced Encryption Standard (AES), which is a digital data encryption standard presented by the National Institute of Standards and Technology (NIST) in 2001. AES has been employed by the US government and has taken the world by storm, replacing early DES released in 1977 [7]. It will be difficult to find industries or services that do not use AES to protect sensitive data from a wide range of products, such as online banking credentials, passwords, and state secrets.

2.1.2.1   Working Mode

The working mode of AES is mainly reflected in the entire process of encrypting a plaintext block into a ciphertext block. AES provides the following five modes: the electronic code book (ECB), cipher block chaining (CBC), counter (CTR), cipher feedback (CFB), and output feedback (OFB) [6].

- *ECB*: Fig. 2.2a gives the workflow of the ECB model, which divides the data to be encrypted into groups of the same size as the encryption key and then encrypts each group with the same key. The advantage of the ECB mode is that it enables parallel computing, but the disadvantage is that the plaintext mode cannot be hidden, so it is vulnerable to active attacks against the plaintext. Therefore, the ECB mode is only suitable for encrypting small messages.
- *CBC*: Fig. 2.2b gives the workflow of the CBC mode, which has better security than the ECB mode and is suitable for transmitting long data packets. The disadvantage is that it is not conducive to parallel computing and requires an initialization vector (IV).
- *CTR*: As illustrated in Fig. 2.2c, in the CTR mode, there is an auto-increment operator. This operator uses the output of the encryption key and the XOR result of the plaintext to obtain the ciphertext, which is equivalent to one encryption.

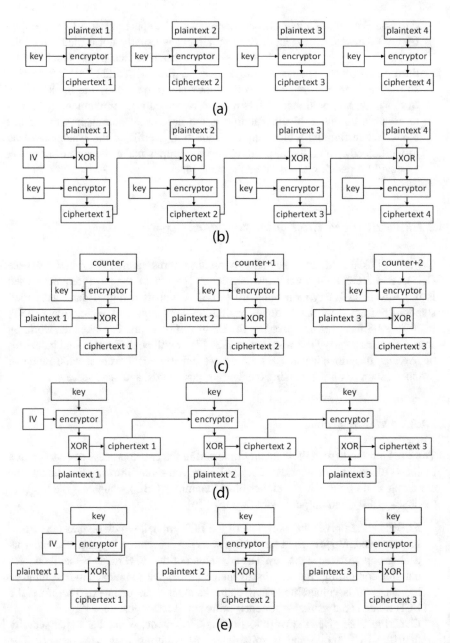

Fig. 2.2 AES working mode. (**a**) ECB mode. (**b**) CBC mode. (**c**) CTR mode. (**d**) CFB mode. (**e**) OFB mode

This encryption method is simple, fast, secure, and reliable, and can be encrypted in parallel.

– *CFB*: As shown in Fig. 2.2d, the CFB mode is akin to the CBC mode; the decryption process of CFB is almost the reverse of the CBC encryption process. The CFB mode hides the plaintext mode, converts the block cipher into a stream mode, and encrypts data smaller than one block. Nevertheless, the CFB mode is not conducive to parallel computing because the damage of one plaintext unit will affect multiple units.

– *OFB*: As demonstrated in Fig. 2.2e, in the OFB mode, block ciphers can be converted to synchronous stream ciphers. The OFB mode generates the key-stream block and then XORs with the plaintext block to get the ciphertext. Like in other stream ciphers, bit flipping in the ciphertext will cause the same bit flipping in the plaintext.

2.1.2.2 AES Encryption

As Fig. 2.3 illustrates [6], the AES encryption algorithm comprises multiple rounds. Each round includes a public permutation (such as byte-substitution, row-shifting, or column-mixing) and a secret conversion adding-round-key, which consists of an XOR operation and a value that depends on the key and the round. In particular, the AES encryption process contains the following steps:

– *Step1*: Dividing data into blocks. Since AES encrypts data in bit blocks, the first step of AES encryption is to divide the plaintext into blocks. Each of its blocks contains a 16-byte column with a 4×4 layout, and since one byte contains 8 bits, the block size is 128 bits ($16 \times 8 = 128$).

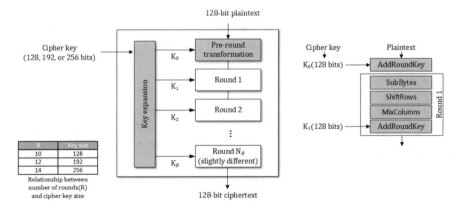

Fig. 2.3 AES encryption keys and rounds

- *Step2*: Key expansion. Taking a 4-word (16-byte) key as input, this step can generate a 44-word (176-byte) linear array, which is enough to provide 4-word round keys for the following round operations.
- *Step3*: Initial round key addition. This is the first round of AES encryption. In this step, the 128 bits of the state and the 128 bits of the round key are XORed bit-wise. This operation is regarded as a column-by-column operation between 4 bytes of the status column and a word of the round key. This step is very important for AES security: If the key is not added in the first round, the first round of public permutation will operate on the plaintext, so there is no cryptographic value involving known plaintext.
- *Step4*: Byte-substitution. This step implements a byte-by-byte replacement for blocks by using the S-box.
- *Step5*: Row-shifting. This step moves the rows of the block obtained during the byte-substitution, cyclically moving the rows to the left. Specifically, the first row remains unchanged, but the second row is shifted to the left by one byte, the third row is shifted by two bytes, and the fourth row is shifted by three bytes.
- *Step6*: Column-mixing. This step uses arithmetic operations on $GF(2^8)$ and performs operations on each column separately. In particular, each byte of the column is mapped to a new value, which is a function of all four bytes in the column. The combination of row-shifting and column-mixing operations makes sure that all output bits depend on all input bits after a few rounds.
- *Step7*: Round key addition. In Step2 and Step6, we achieve the round key and the block, respectively. In this step, we further combine them into one new binary-code block that will be modified later.

The process of AES encryption, subsequently, goes through multiple rounds of byte-substitution, row-shifting, column-mixing, and adding-round-key operations. The round number required is determined by the length of the AES keys:

- Keys of 128 bits: 10-round operations
- Keys of 192 bits: 12-round operations
- Keys of 256 bits: 14-round operations

2.1.2.3 AES Decryption

With the help of reverse encryption, AES ciphertext can be restored to its original state. Especially, the AES decryption algorithm begins with the reverse round key and then reverses each individual operation (byte-substitution, row-shifting, and column-mixing) until it deciphers the original message.

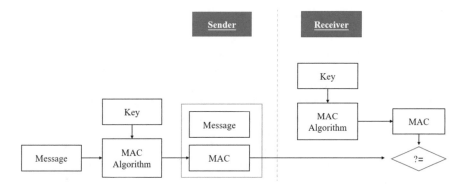

Fig. 2.4 Message authentication code

2.1.3 Message Authentication Code

The message authentication code (MAC) can decide whether the data has been modified due to an attack on data integrity [8, 9]. Specifically, the MAC uses symmetric-key encryption to authenticate messages. The sender and receiver initiate the MAC algorithm by exchanging the symmetric key K. Then, the MAC algorithm calculates the encryption checksum of the message and transmits the calculation result with the message to ensure the authenticity of the message. Figure 2.4 illustrates the process of utilizing MAC for authentication.

Below we further elaborate on the message authentication mechanism based on MAC:

- The MAC algorithm returns the MAC value when given a key and a message. It is worth emphasizing that a MAC function is almost the same as a hash function, which converts an input of any length into an output of a fixed length, except that a key is necessary to calculate the MAC value.
- The sender transmits a message and its corresponding MAC value to the receiver. We assume that messages are delivered in plain text, considering that the focus here is on message integrity and source authentication rather than message confidentiality. If the message needs to be kept secret, we should encrypt it.
- The receiver enters the message and the shared key (K) into the MAC algorithm and recalculates a new MAC value after receiving a message and a MAC value.
- The receiver now verifies the equivalence of the newly calculated MAC with the MAC received from the sender. If so, the recipient ensures that the received message is from the claimed sender; otherwise, the receiver cannot identify whether the transmitted message has been changed or the source has been forged, so the authentication for integrity fails.

2.1.4   Digital Signature

The digital signature may be the most popular public-key cryptographic technique for message verification by binding individuals or entities with digital certificates [10, 11]. The receiver can verify the validity of this binding. Specifically, digital signature algorithms use messages and keys that only the signer knows to generate signature values. In the real world, the recipient of a message needs to make sure that the message belongs to the sender, and the sender cannot deny this. This method is very important in commercial applications because the possibility of inconsistent data exchange is very high. Figure 2.5 depicts the model of digital signature.

Below, we further introduce the entire process of message authentication based on digital signature:

- In a digital signature system, each registered user has a unique public/private key pair, which may be called a signature/verification key pair.
- The signer first uses the given hash function to hash the message. By taking the hash value and the signature key, the signature algorithm produces the signature of the message. Then, the signer links the message to the signature and transmits both to the verifier.
- The verifier inputs the digital signature and the verification key into the verification algorithm, generating a hash value. Then, the verifier uses the same hash function on the incoming message to generate another hash value. Two hash values are compared for verification. The verifier determines the validity of the digital signature based on the comparison result.
- It is worth emphasizing that since the digital signature is generated using a private key owned by only the signer, the signer cannot deny signing the message in the future.

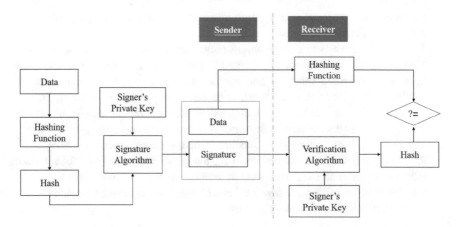

Fig. 2.5   Digital signature

- The signature algorithm is not directly used to sign messages. The hash of the message provides a unique representation of the message, so it is sufficient to sign the hash instead of the message. The main reason is to improve the efficiency of digital signatures. For example, we use RSA as the signature algorithm. Then signing large messages involves computationally expensive and time-consuming modular exponentiation operations. The hash of the message is a relatively small digest of the message, so signing the hash is more efficient than signing the entire message.

The digital signature is a key technology to ensure message security. Besides guaranteeing the non-repudiation of the message, the digital signature also assures the authenticity and integrity of messages. Specifically, the digital signature has the following functions:

- *Data integrity.* If an attacker gains access to the data and makes changes to it, the recipient's digital signature verification will fail. The hash value after changing the data does not match the result of the verification method. Therefore, if the data integrity is compromised, the recipient can safely reject the message.
- *Message authentication.* When the verifier uses the sender's public key to successfully verify the digital signature, he can guarantee that the signature was only generated by the sender and no one else.
- *Non-repudiation.* Assuming only the signer knows the signing key, he can create a unique signature for the message. The recipient can submit the message and its digital signature to a third party as evidence of future disagreements.

2.1.5 Public-Key Infrastructure

Using a pair of keys for security services is an essential feature of the public-key infrastructure (PKI). The key pair comprises two parts: the private key and the public key. Since the public key is publicly available, it can easily be abused. Therefore, a credible infrastructure to be in charge of these keys is necessary [12–14].

The security of a cryptographic system depends on the integrity of key management, which is the process of safely managing keys during the entire key life cycle. If there is no secure key management method, a strong cryptographic system will also be insecure. It is worth noting that encryption algorithms are rarely insecure; more security threats are caused by invalid key management, as shown in Fig. 2.6. Besides, public-key cryptography has two key management requirements: (1) the confidentiality of the private key, which means the private key must be kept secret during the key life cycle, and (2) the correctness of the public key, which can be guaranteed by PKI.

PKI mainly includes the following components: digital certificate, private key, certification authority (CA), registration authority (RA), and certificate management system (CMS):

Fig. 2.6 Key management of PKI

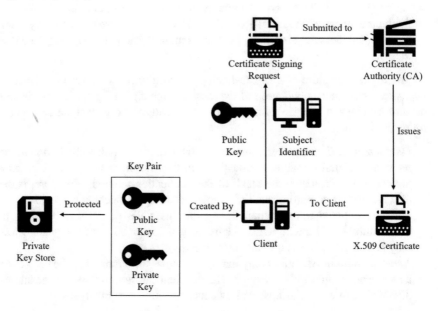

Fig. 2.7 Certification authority

- *Digital certificate.* A digital certificate can be compared with an individual's identification document. People usually use identification documents like a passport or driver's license to prove their identity. In the electronic world, digital certificates perform almost the same functions, but there is an exception: digital certificates are issued to people, as well as to computers, software packages, and anything else that needs to be recognized in the digital world. Digital certificates mainly use the ITU standard X.509, which establishes a standard certificate format for public-key certificates and certificate verification. Therefore, digital certificates are also called X.509 certificates. The CA stores the user's public key and other related information (such as subject name, validity period, purpose, issuer, etc.) in the digital certificate. Then, the CA digitally signs the entire message and includes it in the certificate as a digital signature. The person who wants to ensure the client's public key and related information can use the CA to verify the signature of the certificate. Validation guarantees the certificate's public key pertains to the subject of the certificate.
- *CA.* As shown in Fig. 2.7, the CA acts as a certificate authority by issuing certificates to clients and assisting other users to verify them. The CA is primarily responsible for identifying the client requesting the certificate, as well

as guaranteeing that the contained information in the certificate is accurate and digitally signed. Specifically, the CA has the following important functions:

* Generation of key pairs. The CA generates a public-private key pair independently or jointly with the client.
* Certificate issuance. Once the entity provides the credential required to prove its identity, the CA will issue and sign the certificate to prevent the certificate information from being modified.
* Certificate distribution. The CA must issue certificates so that users can easily obtain them. There are two ways to do this. One option is to issue a certificate in the electronic equivalent of the phone book. Another method is to distribute certificates to anyone who needs them.
* Certificate verification. The CA makes its public key accessible to the public for verifying its signature on the client's digital certificate.
* Certificate revocation. The CA may revoke the issued certificate for various reasons, such as the disclosure of the private key. The CA records all revoked certificates through the certificate revocation list (CRL).

– *RA*. A trusted third party (TTP), called a registration authority (RA), can be entrusted by the CA to check the validity of the certificate applicant's identity, but the RA does not have the function of signing certificates.
– *CMS*. The CA and associated RAs launch CMS that can issue, suspend, renew, and revoke certificates. For legal reasons, it may be necessary to prove certificate status in time, and thus CMS rarely deletes the certificate.
– *Private key*. The client's public key is contained in its certificate, while the corresponding private key is put on the client's computer, but an attacker who has access permission to the computer can simply obtain the private key. In this case, the private key should be stored on a secure portable storage token that requires a password to access.

2.1.6 Identity-Based Cryptography

A major obstacle to the widespread acceptance of public-key cryptography is its heavy reliance on expensive PKIs. Before initiating secure communication, both the sender and receiver must generate a public-private key pair. They then submit a certificate request to the CA and receive a CA-signed certificate, which they can use to authenticate and exchange encrypted messages. This process is time-consuming and unstable. It is also inconvenient for inexperienced people.

In identity-based cryptography (IBC), a party may use its identity string to get directly its public key; a TTP called a private key generator (PKG) can generate the associated private key for the party. Specifically, the PKG first produces a master public-private key pair, and then secretly keeps the master private key and publicly releases the master public key. After a party's public key is calculated based on the master public key and its identity, PKG can use the master private key to distill the party's private key based on its identity. Therefore, IBC simplifies key management, and PKI is no longer indispensable.

2.1.6.1   IBC History

In 1984, Adi Shamir introduced the concept of identity-based cryptography [15]. Its main novelty lies in the use of user identities like phone numbers or email addresses for encryption and signature verification rather than digital certificates. This function does not need to generate and manage user certificates, which greatly reduces the complexity of public-key cryptosystems. After the concept of IBC was proposed by Shamir, he also proposed the construction method of an identity-based signature (IBS), but the problem of identity-based encryption (IBE) has not been solved. Until 2001, Boneh and Franklin [16] proposed the first practical IBE scheme. Since then, IBC has always been a research hotspot in the field of cryptography [17–21].

2.1.6.2   IBC Operation Overview

Before conducting secure communication, PKG must generate its own public-private key pair (pk_{PKG}, sk_{PKG}) and release the public key pk_{PKG}. These keys are called the master public key and the master private key, respectively. Figure 2.8 lists the steps of IBE.

- *Step1*. Alice wants to send a plaintext message M to Bob. She first encrypts M through Bob's identity ID_{Bob} and PKG's public key pk_{PKG}, obtaining the ciphertext C and sending it to Bob. Note that Alice already knows ID_{Bob} and pk_{PKG} before starting the encryption process.
- *Step2*. Bob receives C from Alice. First, Bob sends a private key request to PKG, and PKG verifies the validity of Bob's identity; if so, PKG transmits Bob's private

Fig. 2.8 Identity-based encryption

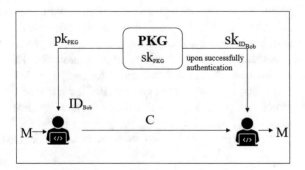

Fig. 2.9 Identity-based
signature

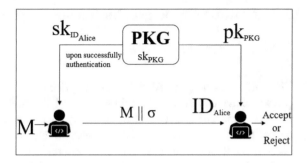

key $sk_{ID_{Bob}}$ to him via a secure channel. For example, to check Bob's identity, PKG delivers a random number to Bob's e-mail address. If the random number is successfully returned, PKG guarantees that the owner of ID_{Bob} is Bob. This random number can be sent through an SSL channel, which provides a secure link for Bob to download his private key.

- *Step3*. Bob uses his private key $sk_{ID_{Bob}}$ to decrypt C to recover the plaintext message M.

As shown in Fig. 2.9, IBS is the mirror image process of IBE, and the steps are given as follows.

- *Step1*. Alice authenticates her identity to PKG and receives her private key, $sk_{ID_{Alice}}$.
- *Step2*. Using her private key $sk_{ID_{Alice}}$, Alice generates a signature δ for M and transmits the message-signature pair (M, δ) to Bob.
- *Step3*. After receiving (M, δ) from Alice, Bob uses Alice's identity ID_{Alice} and PKG's public key pk_{PKG} to check whether δ is a valid signature of M. If so, he returns "Accept"; otherwise "Reject". It is important to mention that Bob does not necessarily provide any certificates to Alice.

2.1.6.3   Revocation in IBC

Although IBC has certain advantages over PKI-based methods, it also has some challenges. In an IBC system, an entity determines its own public key by combining the master public key with its identity. In such a scenario, the public key of the entity cannot be revoked if its identity does not change. The revocation may be a challenging problem for the wide spread of the IBC system. Fortunately, further studies show this problem can be effectively solved by key-evolution techniques [22–25].

2.2 Existing Cryptographic Techniques for ADS-B

2.2.1 Privacy Protection

The public broadcast of ADS-B messages does not consider the confidentiality or privacy of aviation data, which will obviously harm flight safety [26–29]. An ADS-B broadcast may be intercepted by anybody in the receiving range using easy-to-access low-cost receivers. Based on the ADS-B messages received by the ground station, an Internet database has been established to monitor and track aircraft around the world. In this case, you can easily obtain the flight information of any aircraft, such as altitude, current heading, origin/destination, and call sign. This information may not only be used by aircraft enthusiasts but also by malicious attackers [30]. Besides, recent studies have shown that it is comparably effortless to generate and broadcast false ADS-B messages. Injecting the forged ghost aircraft into the surveillance system, in particular, can cause confusion, costly delays, and possibly aircraft accidents.

Notably, a general aviation aircraft needs to broadcast aircraft identification (e.g., a 24-digit ICAO address) and real-time location data for ATC services. However, with the convenient access to air traffic communication and the enhancement of Internet data mining capabilities, anyone can remotely identify and track the aircraft owned by the target user, raising privacy issues for the user. Therefore, it is essential to enable safeguarding the privacy of identity and location data in aircraft ADS-B messages, which may be applied unlawfully for further inferences, for example, aircraft owners' travel intentions and interest profiles [31–33]. Therefore, improving the privacy of ADS-B broadcast messages requires the following security goals:

- *Anonymity.* Unauthorized third parties should not associate the aircraft's ADS-B broadcast with its digital identity.
- *Untraceability.* Unauthorized third parties should not connect two consecutive broadcasts of an aircraft together.
- *Integrity and availability.* The integrity and timeliness of the aircraft's ADS-B messages should be ensured.

To prevent erroneous ADS-B broadcasts, ADS-B messages are required to be protected by cryptographic methods before they are sent. According to the type of key used, cryptographic schemes are divided into two categories: (1) symmetric and (2) asymmetric. The symmetric method uses the same key for encryption and decryption, which is shared by all users. The asymmetric method uses a combination of public and private keys for each user. Messages encrypted with the recipient's public key can only be decrypted with the recipient's private key.

Asymmetric Encryption The asymmetric-key method has significant advantages over the symmetric-key: Alice's public-private key combination cannot be used to forge Bob's asymmetric-key encryption or signature. Therefore, one pair of keys needs to be revoked only with private key leakage. This is in sharp contrast to the symmetric-key method, where a single key leakage may compromise the whole system. While asymmetric-key methods are less computationally efficient than symmetric-key techniques, they may be just as safe. Asymmetric-key methods distribute public-private key pairs through PKI, in which each user has a public-private key pair associated with their identity [13, 34].

The FAA or ICAO may take over the functions of the CA. Asymmetric-key methods can create, distribute, and revoke keys through the use of PKI. Before a flight, aircraft may be loaded with a list of public keys that have altered since the last flight or an outright list of all known public keys. Real-time generation and revocation of keys can be transmitted via satellite or terrestrial data connections, which can be accessed by most commercial flights. Specifically, asymmetric-key encryption schemes may use specific public-key encryption techniques (such as elliptic curve cryptography) to encode ADS-B communications. Typically, the message will be encoded using the public key of the intended recipient, and the recipient will decode it using his private key. Confidentiality is also a by-product of asymmetric-key encryption, as the sender's intended recipient can only decode communications. Notably, if a message is to be delivered to several distinct receivers, it must be sent individually to each recipient. This will require the aircraft to be aware of its neighbors and the ground stations to choose the keys to encrypt each message [35, 36]. Although each aircraft can maintain a public-key list of nearby aircraft and ground stations, the implementation will require repeated broadcasts of the same message using different public keys to encrypt the message for each aircraft and ground station. As the message transmission rate slows, the overhead associated with ADS-B may outweigh its advantages.

Symmetric Encryption Symmetric-key methods are computationally efficient, which are based on the sender and receiver sharing a secret cryptographic key. Without knowledge of the shared secret key, it is computationally impossible to fake or anticipate the encrypted communications produced by symmetric-key algorithms. The secret key cannot be deduced from encrypted communications. Using a single key for encryption and decryption ensures scalability. With ADS-B, the key may be sent out of the band from the controller to the pilot (the controller-pilot data link communication (CPDLC)) and then updated as needed. Notably, the CPDLC allows ATCOs and pilots to interact through a data connection system [37, 38]. During the initial flight plan, the aircraft communicates with the control tower, and once synchronized, it will automatically switch between the Air Route Traffic Control Centers (ARTCCs) on the way. In fact, ADS-B key exchange can be included in the CPDLC login process, allowing secure key exchange. For key management, the ATCO can randomly create a new key as needed and send it to each aircraft equipped with the CPDLC system. From the pilot's view, upgrading the ADS-B encryption key will be like the current radio-changing process of the ATC system.

Besides key distribution for the security of ADS-B communications, we further consider the functioning of the ADS-B system while selecting a suitable symmetric-key algorithm. To defend against known plaintext cryptanalysis assaults, it is essential to diffuse identical patterns in the plaintext into the ciphertext, which means that block ciphers are more suitable for ADS-B scenarios than stream ones. Although there are numerous freely available block ciphers (e.g., DES, AES, and 3DES) [39, 40], these block ciphers require the fixed length of the message block, precluding the use of the ADS-B message format. The length of the ADS-B message is 112 bits, broken down into five fields. For ADS-B, the downlink format field is set to 17, which specifies the communication type. Since this field informs the recipient of the type of message to be processed, it must be left as plain text. The remaining 107 bits are used to store functional data, which can be encrypted without affecting the function [41, 42]. It is important to mention that fixed block ciphers are contingent on the need for accurate encryption of 64-bit or 128-bit blocks. Padding or truncation will apply to messages that do not conform to the specified block size. However, for ADS-B, because the underlying ADS-B architecture is optimized for a 112-bit fixed data length, the message length cannot be increased. In fact, there is a need for an encryption method that can support blocks of any size.

2.2.2 Integrity Assurance

Secure broadcast authentication is an effective method to prevent and/or detect attacks on unidirectional broadcast networks (such as ADS-B). In this section, we describe the current broadcast authentication technologies for wireless sensor networks or vehicular ad hoc networks (VANET) and then evaluate their applicability to ADS-B. In particular, the aim is to keep the openness of ADS-B while providing possible authentication methods. This can be accomplished on a global scale, or it can be done selectively in response to suspicious activity. This reactive authentication can reduce network pressure by requiring security (as well as computational and communication overhead) only when the event is most likely to occur. [43].

MAC The message authentication code is a message digest generated from a message through a MAC generation method (such as keyed-hash MAC or parallelized MAC). The MAC is linked to a longer message, and the message-MAC pair is broadcast at the same time for fast verification. An attacker cannot successfully violate message integrity once a MAC is generated. However, since the plaintext message is still broadcast, the MAC method does not guarantee confidentiality [44, 45]. The MAC mechanism lengthens the transmission time and increases the possibility of ADS-B message interference or overlap during a broadcast. The interference because of supporting MAC on the 1090 MHz channel may shrink the operating capacity of the system, reducing the gain of ADS-B. A lightweight broadcast method is only part of the MAC is sent. Fragmented MAC bits can be added to standard ADS-B messages or transmitted using spare bits in an alternative message format. The delay between the transmission of the initial ADS-B message and the final MAC-based message validation is a defect of the lightweight method.

Digital Signature A digital signature algorithm (for example, DSA or DSS) accepts a message and a user's private key as input and generates a unique digital signature for the message. After receiving the message-signature pair, the receiver can execute the verification method, using the sender's public key to verify the signed message. A successful verification shows that the sender has signed the transmitted message, and no changes have been made. Specifically, ECDSA generates a shorter digital signature for a certain security level given in a series of digital signature algorithms, which makes it easy to adapt to ADS-B [46]. The ECDSA signature of 448 bits is commensurate to the 112-bit symmetric key for security level, so this signature is four times longer than the ADS-B message [47].

In addition, Cook [35] proposed an ADS-B security scheme that combines asymmetric and symmetric encryption: First, PKI registers all ADS-B transponders to the ATC system. Then, asymmetric encryption with distributed symmetric session keys serves as establishing the authenticity and integrity of the transmitted data. Finally, this scheme ensures the security, practicability, and scalability of ADS-B and also minimizes the need for digital signatures to save costs.

2.2.3 Limitations of Existing Cryptographic Methods

The previous section introduces ADS-B security enhancements based on cryptography. We then analyze the limitations of cryptographic methods used in the ADS-B scenario in this section.

2.2.3.1 Limitations of Symmetric Keys

Privacy protection technologies (such as encryption) that provide confidentiality can reduce privacy risks by restricting access to sensitive or personal data. Unfortunately, direct use of these measures may not be suitable for ADS-B communications, because ADS-B broadcasts must remain publicly accessible. Specifically, directly encrypting an entire ADS-B message is not appropriate because it violates the openness of ADS-B and may affect flight safety.

Besides, we will discuss one significant drawback of symmetric-key management. Anyone with the key can create a message, passing cryptographic verification. As a result, a single key leak may jeopardize the whole system. The security of the symmetric-key system entirely depends on the security of the keys. In an ADS-B scenario, it would provide the same encryption keys to all ADS-B participants on a global scale, or at the very least, to aircraft and ground stations within a specified region. Such a large-scale group encryption system, including general aviation, would be deemed highly vulnerable to both internal and external assaults. When keys are dispersed in potentially untrustworthy groups, their life cycles are significantly reduced.

Three methods for key distribution have been suggested: distributing keys in tamper-resistant hardware to all aircraft, distributing keys in tamper-resistant hardware to the selected aircraft, or distributing keys through ATC during a flight. The first is vulnerable to single key leakage and is reliant on the security of tamper-proof. Although the second has been used in both civilian and military contexts, its practicality is debatable. Third, each aircraft will be assigned a unique key for each flight. Before a trip, ATC may issue the flight-specific key and store it in an international database to help contact other aircraft. But this method can easily cause the loss of the entire active key database.

Finally, one of the FAA's goals is to ensure that ADS-B operates on a global scale. Although the FAA does not seem to prohibit explicitly the encryption of civil aviation communications, the FAA has stated that the requirement for encryption of ADS-B signals will unnecessarily restrict international use. Even if international interoperability issues are resolved, the FAA and ICAO may continue to oppose ADS-B encryption because it endangers conventional flight safety: encrypted ADS-B information occasionally cannot be decrypted because of human error or technical failures, increasing the jeopardy of aircraft collisions.

2.2.3.2   Limitations of Asymmetric Keys

The PKI is required to manage and store certificates to ensure the validity and authenticity of public keys. However, with the ever-increasing air traffic, certificate revocation, renewal and storage will emerge as recurrent, costly, and poor in scalability. In addition, the performance of asymmetric cryptographic algorithms (such as digital signatures) is worse than that of symmetric algorithms (such as MAC). Therefore, considering the availability in the ADS-B system, the main concerns of public-key cryptography is how to solve the problems of poor scalability of PKI and the expenditure of digital signature [48].

Some researchers suggested using X.509 certificates as an enhancement of ADS-B communication security [47, 49]. Specifically, public-key certificates and CRLs can be transmitted from the ground to the aircraft through the Aviation Asset Distribution System (AADS).

Despite the technological feasibility of ensure the security of ADS-B communication through public-key cryptography, this will bring huge technical and financial burdens to the aviation industry, since it is very costly to distribute certificates to aircraft and ground stations, protect private keys during transmission and usage, and implement certificate revocation. Specifically, each aircraft must get its public key-private key pair and store the private key securely. It is important to emphasize that each aircraft must also have a list of all other aircraft' public keys to verify their signatures. The size of an X.509 certificate is about 5 KB and, therefore, the size of the public key list of one million aircraft will reach 4G. Besides, the public-key certificate revocation is an important challenge since the voice channel does not intend to cope with it. Although AADS can transmit CRL, it must be changed to interact with aircraft in flight. It can also perform such revocation through the FIS-B system, which is used to convey information and temporary flight restrictions about the airspace. However, because FIS-B broadcasts on UAT, aircraft equipped with 1090ES transponders cannot receive it without adding additional hardware.

2.3 Emerging Cryptographic Techniques for ADS-B

Cryptographic approaches are time-tested to protect wireless communications and can naturally protect ADS-B wireless broadcasts. In this part, especially considering the security requirements of ADS-B, we discuss some promising cryptographic techniques, such as format-preserving encryption (FPE), vector homomorphic encryption (VHE), and timed efficient stream fault-tolerant authentication (TESLA).

2.3.1 Format-Preserving Encryption

ADS-B messages follow a particular format, for example, the 1090ES data link standard specifies a 112-bit ADS-B message, which is composed of five fields: DF (Downlink-Format), CF (Code-Format), AA (Aircraft-Address), Data (ADS-B Data), and PC (Parity-Check). Each field has its meaning. For example, the AA field contains the unique identifier of the aircraft, called the 24-bit ICAO address, which usually requires privacy considerations.

To ensure the privacy of ADS-B communication, we may entirely encrypt a whole ADS-B message. Here, the encrypted location data is no longer publicly available to anyone without keys, potentially jeopardizing flight safety. As an alternative, we may also encrypt only the AA field using classic symmetric block ciphers, e.g., AES. However, AES-like ciphers require additional padding to meet the specified block size, increasing the length of messages and exacerbating the burden on the already congested 1090ES channel.

FPE can encrypt a fixed-length message that does not have typical block sizes (e.g., 64 or 128 bits). FPE has a variety of applications, e.g., the encryption of insurance policy numbers or financial data preserved in older systems [50, 51]. According to [50, 52], the FPE scheme can be constructed based on conventional block ciphers and proved as secure as the underlying cipher. FFX is an instance of FPE implementation, which is derived from AES. FFX is an abbreviation for a Feistel-based encryption method that preserves message format, with the X denoting multiple parameter sets. Compared with encryption schemes that require a fixed message block size, such as AES, FFX has a unique advantage in the ADS-B environment, in that it can encrypt messages of any length (e.g., the 112-bit ADS-B message). As a result, we can use FFX to encrypt the AA field in the ADS-B message with no additional padding to account for the fixed block length, which is yet consistent with ADS-B openness. Therefore, we introduce FFX to maintain comparability with existing ADS-B protocols.

2.3.1.1 An Illustration of Encrypting ADS-B Messages Using FFX

Below, we use FFX to encrypt ADS-B messages while not violating the openness of ADS-B systems. First, Table 2.1 gives notations commonly used in FFX and Table 2.2 defines FFX parameters including the key, tweak, radix, message length, number of Feistel rounds and split number.

Using FFX, we propose an ADS-B message encryption scheme F, consisting of three PPT algorithms: $F.KG$, $F.Enc$, and $F.Dec$. Before elaborating on this proposal, we first give a brief description on its main idea, as shown in Fig. 2.10. To begin, the ATCO distributes a key k to aircraft through a secure channel. The broadcaster then invokes $F.Enc$ to encrypt an ADS-B message x to be broadcast.

Table 2.1 Notations in FPE

Notation	Meaning
K	Space of keys
Ω	Space of formats
T	Space of tweaks
X	Space of messages over an alphabet $Chars = \{0, 1, \cdots, radix - 1\}$
$Chars^*$	Set of strings with arbitrary length over $Chars$
$|x|$	Bit length of x

Table 2.2 Parameters of FFX

Parameter	Description
$radix$	$radix \geq 2$
$Length$	Set of permitted message lengths
$method$	Feistel method, either 1 or 2
$split(n)$	Function of taking a permitted length $n \in Length$ and returning $1 \leq split(n) \leq n/2$
$rnds(n)$	Function of taking a permitted length $n \in Length$ and returning an even number $rnds(n)$
$Addition$	Addition operator ($\boxplus$), either 0 (character-wise) or 1 (block-wise)
$F_K(n, t, i, B)$	Function of taking a key $k \in K$, a permitted length $n \in Length$, a tweak $t \in T$,
	a round number $i \in 0, \cdots, rnds(n) - 1$, and a string $B \in Chars^*$,
	and returning a string $F_K(n, t, i, B) \in Chars^*$
	If $method = 1$ or i is even then
	$|B| = n - split(n)$ and $|F_K(n, t, i, B)| = split(n)$
	If $method = 2$ or i is odd then
	$|B| = split(n)$ and $|F_K(n, t, i, B)| = n - split(n)$

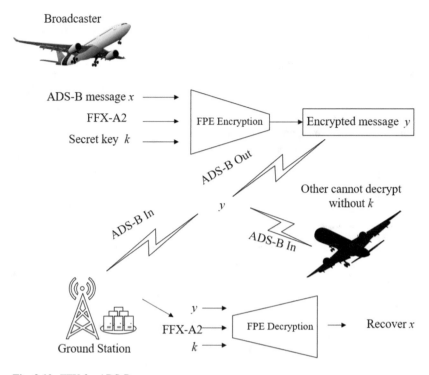

Broadcaster

ADS-B message x ⟶

FFX-A2 ⟶ FPE Encryption ⟶ Encrypted message y

Secret key k ⟶

ADS-B Out

y

Other cannot decrypt
without k

ADS-B In

ADS-B In

Ground Station

y ⟶ FFX-A2 ⟶ FPE Decryption ⟶ Recover x

k ⟶

Fig. 2.10 FFX for ADS-B messages

Recipients with the key k can run $F.Dec$ to recover x. Other non-participants cannot retrieve x without k.

We describe the proposed scheme below. Please note that the ADS-B message of 1090ES contains 112 bits and breaks down to five fields. Although the 5-bit DF field shows the message type and should remain unencrypted, we can encrypt the next 107 bits of stored function data without affecting the function.

- $F.KG$. The ATCO plays a role of TTP initializing the whole system in the following steps:

  - $Step1$. Specify the binary string format as $\{0, 1\}^{107}$.
  - $Step2$. Choose the parameter set of FFX-A2 according to Table 2.3, in which AES-128 is used.
  - $Step3$. Generate randomly a key k and then transmit it to each aircraft.

- $F.Enc$. The broadcaster encrypts an ADS-B message x with the key k and broadcasts the corresponding ciphertext y as shown in Fig. 2.11.
- $F.Dec$. Upon the receipt of a ciphertext y, each receiver with k runs the decryption algorithm to recover the corresponding message x as shown in Fig. 2.12.

Table 2.3 FFX-A2 parameters

Parameter	A2
radix	Value $= 2$ meaning $Chars = \{0, 1\}$
Length	Value $= 117$ meaning that the length of *plaintexts* (or ciphertexts) is no more than 117 bits
keys	$\{0, 1\}^{128}$
tweak	$Byte^{\leq M}$, $M = 2^{64} - 1$
addition	Value $= 0$ meaning that the character-wise *addition* is used
method	Alternative Feistel is used
split	Maximally balance is used
rnds	Number of rounds is 12 or more
F	Round function is AES-based

- **Algorithm Encrypt (k, t, x)**

- **Input**: a key k, a tweak t and a plaintext x.
- **Output**: a ciphertext y.
- **Steps**:

    1.   if $k \notin K$ or $t \notin T$ or $x \notin X$ or $|x| \notin Lengh$ then return $\perp$

    2. $n \leftarrow |x|$; $l \leftarrow split(n)$; $r \leftarrow rnds(n)$

    3.   if $method = 1$ then
           for $i \leftarrow 0$ to $r - 1$ do
              $A \leftarrow x[1...l]$; $B \leftarrow x[l+1...n]$
              $C \leftarrow A \boxplus F_K(n, t, i, B)$
              $x \leftarrow B \| C$
           return x
        end if

    4.   if $method = 2$ then
           $A \leftarrow x[1...l]$; $B \leftarrow x[l+1...n]$
           for $i \leftarrow 0$ to $r - 1$ do
              $C \leftarrow A \boxplus F_K(n, t, i, B)$
              $A \leftarrow B$; $B \leftarrow C$
           return $A \| B$
        end if

Fig. 2.11 FFX encryption algorithm

Our ADS-B broadcast encryption scheme builds its security on the underlying FFX. As mentioned in [52], FFX can prove secure, so our solution is also secure. Subsequently, we assess the performance of our scheme. Based on the MIRACL cryptographic library [53], we then implement our scheme. Finally, we simulate

---

- **Algorithm Decrypt(k, t, x)**

- **Input**: a key k, a tweak t and a ciphertext y.
- **Output**: a ciphertext y.
- **Steps**:

  1.   if $k \notin K$ or $t \notin T$ or $x \notin X$ or $|x| \notin Lengh$ then return $\bot$
  2. $n \leftarrow |x|$; $A \leftarrow split(n)$; $r \leftarrow rnds(n)$
  3.   if $method = 1$ then
         for $i \leftarrow 0$ to $r - 1$ do
            $A \leftarrow x\,[1...l]$; $B \leftarrow x\,[l+1...\,n]$
            $C \leftarrow A \boxplus F_K\,(n, t, i, B)$
            $x \leftarrow B \| C$
         return x
       end if
  4.   if $method = 2$ then
         $A \leftarrow x\,[1...l]$; $B \leftarrow x\,[l+1...\,n]$
         for $i \leftarrow 0$ to $r - 1$ do
            $C \leftarrow A \boxplus F_K\,(n, t, i, B)$
            $A \leftarrow B$; $B \leftarrow C$
         return $A \| B$
       end if

Fig. 2.12 FFX decryption algorithm

running on the ARMv7 microprocessor and 1GB RAM, and the performance of this encryption scheme is summarized in Table 2.4.

2.3.2   Vector Homomorphic Encryption

The VHE is a new homomorphic encryption (HE) method that can encrypt a plaintext vector in a batch mode. To protect aircraft location privacy, we introduce VHE to encrypt aircraft positional fingerprint vectors efficiently [54]. Although Bogos et al. have demonstrated security issues on the original VHE [55], an improved VHE has been proposed to fix such security flaws.

Table 2.4 FFX encryption
time for ADS-B messages

CPU frequency (MHz)	Encryption time (us)
200	498.8047
400	352.6445
600	290.7148
800	231.4375
1200	167.1133
1400	124.9395

2.3.2.1 Original Scheme

The original VHE scheme was proposed in [54] consisting of six PPT algorithms as
follows.

- *K G (Key generation).* Taking a security level λ as input, first choose six integers
  $l, m, n, p, q, w \in \mathbb{Z}$ satisfying $m < n$, $q \gg p$, $w(p - 1) < q$, and an error
  distribution χ on $\mathbb{Z}_q$; then generate a matrix $S = [I, T]$, where I is an $m \times m$
  identity matrix and T is a $m \times (n - m)$ random matrix; finally keep S as the
  secret key and publish $Param = (l, m, n, p, q, w, \chi)$ as the pubic parameter.
- *Enc (Encryption).* Taking a message vector $x \in \mathbb{Z}_p^m$ and the secret key $S \in \mathbb{Z}^{m \times n}$
  as input, choose an error vector $e \in \chi^n$ and calculate a ciphertext $c \in \mathbb{Z}_q^n$ as

$$Sc = wx + e \tag{2.1}$$

- *Dec (Decryption).* Taking a ciphertext vector $c \in \mathbb{Z}_q^n$ and the secret key $S \in
  \mathbb{Z}^{m \times n}$ as input, retrieve a message $x \in \mathbb{Z}_p^m$ as

$$x = \left\lceil \frac{Sc}{w} \right\rfloor_q \tag{2.2}$$

- *K S (Key switching).* Convert an old key/ciphertext pair (S_{old}, c_{old}) to a new pair
  (S_{new}, c_{new}) with the same message x. Concretely, we have $c_{new} = Mc_{old}$ and
  $S_{new}c_{new} = S_{old}c_{old} = wx + e$, where M is a key-switching matrix. Especially,
  we have

$$(M, S_{new}) \leftarrow VHE.KS(S_{old}) \tag{2.3}$$

- *Add (Addition).* Given two ciphertexts c_1, c_2 that, respectively, correspond to
  two plaintexts x_1, x_2 under the same secret key S, we have

$$S(c_1 + c_2) = w(x_1 + x_2) + (e_1 + e_2) \tag{2.4}$$

- *LT (Linear transformation).* Given a ciphertext c that corresponds to a plaintext
  x under the secret key S, we have the linear transformation Gx as

$$(GS)c = w(Gx) + Ge \tag{2.5}$$

in which G is a linear-transforming matrix. Here we may consider that c is a ciphertext of Gx under GS.

- IP *(Inner product)*. Given two ciphertexts c_1, c_2 that, respectively, correspond to two plaintexts x_1, x_2 under the same key S, and a vectorization function $vec(\cdot)$, we have the inner product $x_1^T x_2$ as

$$vec(S_1^T S_2)^T \left[\frac{vec(c_1 c_2^T)}{w} \right] = w(x_1^T x_2) + e \tag{2.6}$$

2.3.2.2   Improved Scheme

The original VHE [54] is insecure because an attacker can extract the secret key or the plaintext vector from the key-switching matrix or the ciphertext, respectively [55]. Binary strings are used to represent the elements of vectors and matrices in the original VHE. Therefore, bit-by-bit operations on high-dimension vectors incur high computational and communication costs. For example, the public-key size required to encrypt a 500-dimensional vector is 120 MB, and the maximum encryption time is 27 s. As a result, the original VHE is unsuitable with ADS-B communications considering low-bandwidth data links and resource-constrained avionics.

In this section, we propose an improved VHE scheme to enhance the security and performance of VHE. To be more precise, we first construct an invertible matrix to generate a new key-switching matrix, which can prevent secret-key recovery. We then insert an error into a message for obfuscation during message encryption. In doing so, the adversary cannot recover the secret key or the message from the key-switching matrix or the encrypted vector, receptively. As a result, the improved VHE scheme guarantees indistinguishability under chosen plaintext attacks (IND-CPA), i.e., semantic security. Since our VHE does not use the binary representation of vectors or matrices, it offers the benefit of a short public-key size and a low encryption time cost.

`2.3.2.3   Scheme Description

We begin by introducing two new algorithms, *GenInv* and *Trans*, which are utilized to generate the parameters of the improved VHE scheme.

In *GenInv*, I_1 and I_2 are generated satisfying $I_1 I_2 = I$, where I is an identity matrix. Note that in *GenInv*, the large loop value of k would result in the good randomness of I_1 and I_2.

In *Trans*, n and n' denote the dimensions of S_{old} and S, respectively. Provided with an old ciphertext c_{old} and its associated secret key S_{old}, *Trans* outputs S and

---

Algorithm 1 *GenInv*

1: **Input:** n: Dimension of ciphertext vectors
2: **Output:** $I_1, I_2 \in \mathbb{Z}_q^{n \times n}$
3: **for** $i = 1$ to k **do**
4:      Let $s, d \leftarrow \chi, s \neq d, o \leftarrow \{-1, 0, 1\}$
5:      **if** $o = 0$ **then**
6:          Column swap $I_1[s]$ and $I_1[d]$
7:          Row swap $I_2[d]$ and $I_2[s]$
8:      **else if** $o \neq 0$ **then**
9:          $I_1[s] \leftarrow I_1[s] - o \cdot I_1[d]$
10:         $I_2[d] \leftarrow I_2[d] + o \cdot I_2[s]$
11:      **end if**
12: **end for**

Algorithm 2 *Trans*

1: **Input:** $S_{old} \in \mathbb{Z}_q^{m \times n}$: Old secret-key matrix
2: **Output:** $S \in \mathbb{Z}_q^{m \times n'}$: New secret-key matrix;
     $M \in \mathbb{Z}_q^{n' \times m}$: Key-Switching matrix
3: Get $P_s, P_m \leftarrow GenInv(n')$
4: Get $T \leftarrow \chi^{m \times (n'-m)}$ and $A \leftarrow \chi^{(n'-m) \times n}$
5: Calculate $S_t = [I, T] \in \mathbb{Z}^{m \times n'}$, in which I is an identity matrix, and $M_t = \begin{bmatrix} S_{old} - TA \\ A \end{bmatrix} \in$
     $\mathbb{Z}_q^{n' \times m}$
6: Calculate $S = S_t P_s$ and $M = P_m M_t$

M. In this case, we calculate a new ciphertext as $c = M c_{old}$, and the correctness can be ensured as follows.

$$Sc = SM c_{old}$$
$$= S_t P_s P_m M_t c_{old}$$
$$= [I, T] \begin{bmatrix} S_{old} - TA \\ A \end{bmatrix} c_{old}$$
$$= S_{old} c_{old}$$

Then, we propose an improved VHE scheme, including a quadruple of algorithms as $VHE = (Setup, KG, Enc, Dec)$.

- $Setup(\lambda)$. Taking a security level λ, choose $m, n, p, q, w \in \mathbb{Z}$ satisfying $m < n$ and $q \gg p$, and set a discrete Gaussian distribution χ on $\mathbb{Z}_q$ with standard deviation δ. To guarantee practical security, the setting of q is based on λ. Then publish $Param = (m, n, p, q, w, \chi)$ as the system parameter.
- $KG(Param)$. Taking the system parameter $Param$, running Alg. 1 to produce $(S, M) \leftarrow Trans(wI)$ with an identity matrix $I \in \mathbb{Z}^{m \times m}$. Keep privately the

secret key $S \in \mathbb{Z}_q^{m \times n}$, and release the encryption key $M \in \mathbb{Z}_q^{n \times m}$. According to Alg. 12, we have $SM = wI$.

- $Enc(x, M)$. Taking the message $x \in \mathbb{Z}_p^m$ and the key-switching matrix $M \in \mathbb{Z}_q^{n \times m}$, firstly choose a small error $e \leftarrow \chi^n$ and then calculate the ciphertext $c = Mx + e$.

- $Dec(c, S)$. Taking the ciphertext $c \in \mathbb{Z}_q^n$ and the secret key $S \in \mathbb{Z}_q^{m \times n}$, recover the message $x \in \mathbb{Z}_p^m$. The decryption correctness can be assured as

$$x = \left\lceil \frac{Sc}{w} \right\rfloor_q = \left\lceil \frac{S(Mx + e)}{w} \right\rfloor_q = \left\lceil x + \frac{Se}{w} \right\rfloor_q$$

To guarantee decryption correctness, $|Se| < \frac{w}{2}$ should hold, that is, $n|S||e| < \frac{w}{2}$. Let E be the upper bound of $|e|$, then we have

$$E < \frac{w}{2n|S|}$$

Our improvement has little effect on the homomorphic operations, which means the improved scheme supports all homomorphic operations in the original scheme, including addition, linear transformation, and inner product. However, the conditions for guaranteeing homomorphism would accordingly change as below illustrated.

Addition. Given $c_a = Enc(M, x_a)$, and $c_b = Enc(M, x_b)$, there should be $Dec(c_a + c_b) = x_a + x_b$ for homomorphic addition as

$$Dec(S, c_a + c_b) = \left\lceil \frac{S(c_a + c_b)}{w} \right\rfloor_q$$

$$= \left\lceil \frac{SM(x_a + x_b) + S(e_a + e_b)}{w} \right\rfloor_q$$

$$= \left\lceil SM(x_a + x_b) + \frac{S(e_a + e_b)}{w} \right\rfloor_q$$

$$= x_a + x_b$$

As a consequence, we have the following condition satisfying the homomorphism:

$$\left| \frac{S(e_1 + e_2)}{w} \right| < \frac{1}{2}$$

Linear transformation. Given a matrix G and a vector x, the linear transformation Gx can be calculated in the following way:

$$Gx = GDec(S, c) = G \left\lceil \frac{Sc}{w} \right\rfloor_q = \left\lceil \frac{GSc}{w} \right\rfloor_q$$

Vector c can be treated as the ciphertext of Gx with the corresponding secret key GS. Taking GS as the input of the algorithm $Trans$, the output would be a new secret key S' and a transformation matrix M'. Further, the ciphertext of Gx can be decrypted with S' as

$$\begin{aligned} Dec(S', c') &= \left\lceil \frac{S'c'}{w} \right\rfloor_q \\ &= \left\lceil \frac{GSc + S'e'}{w} \right\rfloor_q \\ &= \left\lceil \frac{G(wx + Se) + S'e'}{w} \right\rfloor_q \\ &= \left\lceil Gx + \frac{GSe + S'e'}{w} \right\rfloor_q \end{aligned}$$

Consequently, we have the following condition satisfying the homomorphism:

$$|GSe + S'e'| < \frac{w}{2}.$$

Inner product. Given two vectors x_1, x_2 and a weight matrix H_j, for the weighted inner product we should compute $x_1^T H_j x_2$ To compute it with ciphertext, we need to use the following equation:

$$(S_1 c_1)^T H_j (S_2 c_2) = vec(S_1^T H_j S_2) \cdot (c_1 c_2^T)$$

Symbol $vec(A)$ stands for a vector consists of all the columns of matrix A. Expand the left part of this equation, we have

$$\begin{aligned} (S_1 c_1)^T H_j (S_2 c_2) &= (wx_1 + S_1 e_1)^T H_j (wx_2 + S_2 e_2) \\ &= w^2 x_2^T H_j x_2 + wx_1^T H_j S_2 e_2 \\ &\quad + (S_1 e_1)^T H_j wx_2 + (S_1 e_1)^T H_j (S_2 e_2) \end{aligned}$$

The jth row of new secret key S' should be $vec(S_1^T H_j S_2^T)$. Its corresponding ciphertext should be $c' = \left\lceil \frac{vec(c_1 c_2^T)}{w} \right\rfloor_q$. However, the inner product operation would introduce much noise into the ciphertext, which limited the depth of homomorphic evaluations. The way to improve depth of homomorphic evaluation is choosing a w

which is large enough to bear all the noises. We do not recommend the inner product evaluation to use in applications.

2.3.2.4 Security

We build the security of the improved VHE scheme upon the learning-with-error (LWE) problem [56]. We then present the proof framework as follows.

Theorem 2.1 *Given the intractability of LWE, our VHE scheme is semantically secure.*

Proof To prove our proposal's semantic security, we firstly prove that it is one-way secure, i.e., an adversary is required to return the correct message x provided with an encrypted message $c = Mx + e$. This is equivalent to solving LWE, namely extracting x from $c \approx Mx$ for a random M. It also means that if random, our proposal's one-way security can be reduced to LWE. We then analyze the randomness of M with $M = P_m M_t$. Obviously, this can be partly guaranteed if M_t has a large size, which is determined by the size of T and A. At the same time, P_m is also required to be random. We can infer each row in P_m has no dependence on other $n - 1$ rows. To further explain this point, let P_1 stand for the probability that a row in P_m is independent of other rows, and we have

$$
P_1 = \left(\frac{\binom{n-1}{2}}{\binom{n}{2}} \right)^k = \left(1 - \frac{2}{n} \right)^k
$$

Let P_2 denote the probability that there is only a row-swap operation for a row in P_m, and we have

$$
P_2 = \frac{1}{3} \left(\frac{\binom{n-1}{2}}{\binom{n}{2}} \right)^{k-1} \frac{\binom{n-1}{1}}{\binom{n}{2}} = \frac{2}{3} \left(1 - \frac{2}{n} \right)^{k-1}
$$

Obviously, if the loop value k is sufficiently large considering the dimension n of ciphertext vectors, then P_1 and P_2 can be neglected. In this case, both P_m and M_t have good randomness. It means that our VHE is one-way secure with respect to LWE hardness.

We then discuss the security of IND-CPA, where an adversary is able to issue polynomial queries to the encryption oracle that is then required to answer with a correct ciphertext of any adaptively chosen messages. That is, given a challenge c that is the ciphertext of a or b, the adversary is required to tell which message of a or

b corresponds to c. First, IND-CPA cannot be accomplished if an adversary violates one-way security. Fortunately, our VHE is one-way secure. Second, by utilizing homomorphic operations supported by our VHE scheme, given c, distinguishing a and b is equivalent to distinguishing $a - a$ and $a - b$ and it is also commensurate to distinguishing the zero vector $\mathbf{0}$ and the non-zero vector x. Since $e_0, e_1 \leftarrow \chi^n$, it is indistinguishable between c_0 and c_1. Therefore, our VHE satisfies IND-CPA, i.e., semantic security.

2.3.2.5   Performance Comparison

We provide a performance comparison between our VHE and the original VHE by conducting simulation tests, which are executed on a desktop with the CPU of Intel i5-8250U and the RAM of 8G running the operating system of Windows 10. To ensure the correctness of decryption, $w = 2^{30}$ and $E = 200$ are set. Figure 2.13a,b illustrate the comparative findings. As can be observed, the higher the dimension of plain vector is, the more space for public-key storage our VHE saves and the faster our VHE encrypts data in comparison to the original VHE. Below, we provide a short analysis of the comparison result.

The original and the improved have key-switching matrices as $M_o = \begin{bmatrix} (wI)^* - TA + E \\ A \end{bmatrix} \in \mathbb{Z}_q^{n'l \times m}$, and $M_t = \begin{bmatrix} wI - TA \\ A \end{bmatrix} \in \mathbb{Z}_q^{n' \times m}$, respectively. The former, in binary, represents wI with the size of $n'l \times m$; the latter, directly in the unit of each element, represents wI with the size of $n' \times m$. The former conducts encryption as $c = Mx$ where the public key M is exactly M_o; the latter performs encryption as $c = Mx + e$ where the public key can be calculated as $M = P_m M_t$. Since the latter does not carry out bit decomposition and thus the binary parameter l is unnecessary, the improved VHE has the merits of the short public key and the few encryption time.

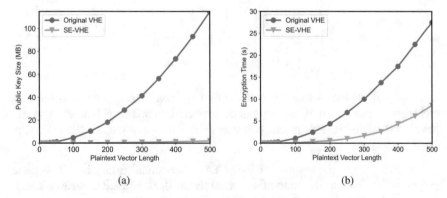

(a)   (b)

Fig. 2.13 Performance of improved VHE. (**a**) Public key size. (**b**) Encryption time

2.3.3   TESLA Authentication

The TESLA technique is a symmetric cryptography-based broadcast authentication mechanism. It generates private keys using one-way hash chains to guarantee source authentication [57]. As shown in Fig. 2.14, to validate the data packet P_i, the sender commits to the key-value K_i by delivering $H(K_i)$ in the packet P_{i-1}. The sender has exclusive knowledge of K_i, which is used to compute the MAC on the packet P_i. The sender reveals the key-value K_i in the data packet P_{i+1} after all receivers have received P_i. The receiver checks whether the received key matches the commitment and whether the MAC value of P_i calculated using the received key matches the received one. If both checks succeed, P_i is considered authentic. Note that P_i includes a commitment for the next key-value pair, K_{i+1}. The first packet is signed using a digital signature method. If P_{i-1} is lost, the authenticity of P_i and all future data packets cannot be confirmed due to the loss of the commitment to K_i. Equivalently, if P_{i+1} is lost, K_i is also destroyed, and the validity of P_i and future data packets cannot be also validated.

Aiming at the issue above, a solution was suggested by creating the sequence of keys K_i in an iterative manner of a pseudo-random function F for some given value as shown in Fig. 2.15. Let $F^0(x) = x$, we denote v consecutive F functions by $F^v(x) = F^{v-1}(F(x))$. In this case, the sender randomly chooses an initial key-value K_n and then pre-calculates n key values, $K_i = F^{n-i}(K_n)$, $i = 0, \cdots, n$. Such the sequence of key values $K_0, \cdots, K_{n-1}$ is called a key chain. As F is a one-way function, provided with only K_i the adversary cannot calculate any K_j for $j > i$. By $K_i' \leftarrow F'(K_i)$, we further derive K_i', which is utilized to authenticate P_i. Nevertheless, for $j < i$, any K_j can be calculated from the received K_i, which means that upon the receipt of P_j, any subsequent P_i allows verifying the authenticity of P_j. Note that a target-collision-resistant pseudo-random function f can be used to implement $F(K_i) = f_{K_i}(0)$ and $F'(K_i) = f'_{K_i}(1)$. Besides, this solution is secure assuming the recipient is capable of deciding the secure arrival of a given packet, that is, the sender has sent the packet before the disclosure of the corresponding key. In this case, only when the recipients have received P_i, the sender can send P_{i+1}, which greatly affects packet rates. To cope with the problem, K_i of P_i can be disclosed in P_{i+d} rather than P_{i+1} [57]. Figure 2.15 depicts

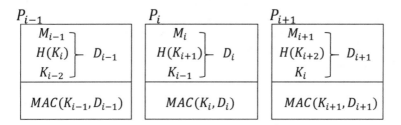

Fig. 2.14 Basic TESLA

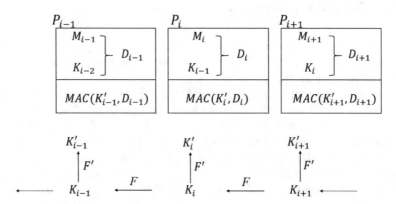

Fig. 2.15 Tolerating packet loss using TESLA

another assumption that packet scheduling is unchanged or foreseeable and that each receiver knows accurately when each packet is sent, which obviously limits the flexibility of the transmission. The issue of dynamic transmission rate can be solved by selecting the MAC key and disclosing one based on the time period, namely using K_i to authenticate all packets in period i. The Internet Engineering Task Force (IETF) adopted the final version of this solution as an Internet standard. Interested readers please refer to [58–60] for more details.

Although the TESLA protocol can achieve integrity for ADS-B messages, it may also introduce transmission delay since nodes are required to buffer messages. As a wireless-sensor-network version of TESLA, μTESLA [61] is introduced that the nodes in the network maintain loose time synchronization, with each node having an upper constraint on the maximum clock synchronization error. As previously stated, typical asymmetric PKI techniques have a large communication cost, casting doubt on their efficacy as security solutions for bandwidth-constrained applications such as the ADS-B network. The μTESLA protocol solves this problem by using asymmetric-key encryption and delaying the publication of symmetric keys, thus bringing in an efficient broadcast authentication mechanism. When interference and bandwidth limits on ADS-B data links are critical, μTESLA may be an option for ADS-B security.

Nonetheless, implementing μTESLA on ADS-B presents two challenges. The first issue is that the ADS-B protocol and message structure must be modified to accommodate the GPS timestamp data. The second problem is that μTESLA must be reinitialized to validate the identity of a node, making it vulnerable to memory-based DoS attacks. Regardless of these drawbacks, μTESLA provides an adequate security solution for ADS-B integration.

2.3.4   Location-Privacy Measurement

Although ADS-B technology provides many benefits such as flight safety, it has location-privacy issues that may include people's travel destinations, commercial secrets, and so on. The *Plane Finder AR* App, for instance, can use ADS-B signals to achieve precise aircraft monitoring. As a result, it is imperative to design a method to preserve ADS-B location privacy. Sampigethaya et al. [62] first suggested a general aviation location-privacy-protection scheme called the "random silence period." Here, the spatial-temporal uncertainty is introduced by updating the aircraft identifiers but not spreading them within a short time, i.e., the maximum silence period. During this period, the target aircraft is mixed with other aircraft in a certain area, thereby increasing the aircraft's recognition difficulty. In addition, this strategy also ensures geographic privacy by assuming that each member of the anonymity set may also be a candidate for the target. Nevertheless, the direct use of the random silence period is only suitable for an ideal condition. The attacker may still practically monitor a target aircraft by predicting the trajectory of the target in advance, thereby reducing uncertainty. In this part, we provide a new correlation tracking method in which an adversary can give a non-uniform probability distribution to the target anonymous set. Therefore, our proposal is more accurate and practical than traditional tracking methods.

2.3.4.1   Proposed Approach

We first assess the level of location privacy of the target aircraft. Entropy is a well-known metric for measuring uncertainty and can be used here to quantify the location privacy of anonymity sets that contain the target aircraft. We use S and $|S|$ to denote a target anonymity set and its size, respectively. For the target T, let the probability $\{i = T | i \in S\}$ be P_i, and we have $\sum_{i=1}^{|S|} P_i = 1$. Then the entropy of S is defined as

$$H(S) = -\sum_{i=1}^{|S|} P_i log_2 P_i \tag{2.7}$$

We further define the reachable area of target T as a bounded area, where T reappears after its identifier is updated [62]. If the target identifier is updated during random silence, there exist some factors related to the determination of the reachable area, including the allowed movement direction, horizontal and vertical minimum separations, $hsep_{min}$, $vsep_{min}$, respectively, the known achievable speed range $[S_{min}, S_{max}]$, and the update period which is between a minimum and maximum silence periods $[SP_{min}, SP_{max}]$. The target anonymity set is comprised of all aircraft that reappear in the reachable area of the target after their identifiers are updated [63].

Fig. 2.16 Simple tracking
method for location privacy

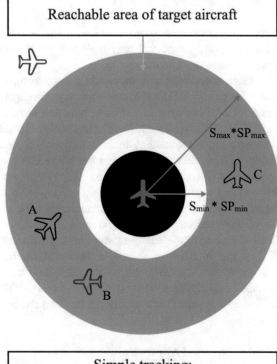

Simple Tracking. For this method, the attacker considers that each member of the anonymity set has the same probability of being a candidate as the target, so the attacker will casually select a member to be the target [64]. Then we have

$$P_i = \frac{1}{E\{|S|\}} \tag{2.8}$$

Figure 2.16 shows that the target anonymity set is at best composed of these aircraft that are located in the circular area centered the position, where the target enters the random silence period. Supposing these aircraft are uniformly distributed in the spatial Poisson process, with a density of ρ in the airspace, and the boundary of their distribution area A_r is the expectation value $E\{|S|\}$ of the anonymity set size at each update [64], then we have

$$1 \le E\{|S|\} \le \frac{\rho A_r}{1 - e^{-\rho A_r}}, \tag{2.9}$$

and

$$A_r = \pi \left[2 \left(S_{max} S P_{max} \right)^2 - hsep_{min}^2 \right] \tag{2.10}$$

Here, for the location-privacy level achievable at each pseudonym update, by using the derived upper bound, the theoretical maximum can be determined as

$$
\begin{aligned}
H(S) &= -\sum_{i=1}^{|S|} P_i log_2 P_i \\
&= -\sum_{i=1}^{|S|} \frac{1}{E\{|S_{A_r}|\}} log_2 \frac{1}{E\{|S_{A_r}|\}} \\
&= log_2 E\{|S_{A_r}|\}
\end{aligned}
\tag{2.11}
$$

Further, we have the expectation size of the anonymity set of a target as

$$
\begin{aligned}
E\{|S_{A_r}|\} &= E\{ \upsilon\{A_r\} | \upsilon\{A_r\} \geq 1 \} \\
&= \frac{E\{\upsilon\{A_r\}\}}{1 - \Pr\{\upsilon\{A_r\} = 0\}} \\
&= \frac{\rho A_r}{1 - e^{-\rho A_r}}
\end{aligned}
\tag{2.12}
$$

Hence, the bounds for entropy are

$$0 \leq H(S) \leq log_2 \left(\frac{\rho \pi R}{1 - e^{-\rho \pi R}} \right) \tag{2.13}$$

As a consequence, we can use the upper bound to obtain the theoretical maximum of the location-privacy level.

Correlation Tracking For this method, the attacker can accredit a non-uniform probability distribution to the target anonymity set and calculate P_i in the following four steps.

- *Step1.* We use the same method as simple tracking to determine the target anonymity set S.
- *Step2.* As illustrated in Fig. 2.17, we first determine A_r, with respect to the target T's last-known position L_{known}, velocity SP_{known}, and direction DR_{known} at time t. Assuming T does not change its velocity and direction during the random silence period, an attacker estimates T's position L_{known1} in A_r at a future time $speriod_{min} + bperiod$, in which $speriod_{min}$ is the minimum silent period and $bperiod$ is the broadcast period. Hence we have $t_i = speriod_{min} + $

Fig. 2.17 Correlation
tracking method for location
privacy

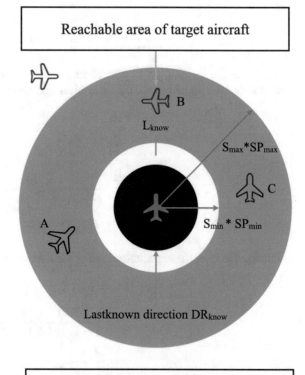

$(i - 1)\,bperiod$, where $t_i \leq speriod_{max}$. As a consequence, the attacker obtains
up to n estimated target positions $\{L_{knowni}\}_{i=1}^{n}$ at time $\{t + t_i\}_{i=1}^{n}$.

– *Step3.* After each update, the attacker selects the aircraft that seems closest to
L_{known}. After each broadcast cycle, the correlation trace is reiterated in A_r before
the maximum silence period $period_{max}$ is reached.

– *Step4.* Assuming that there are n aircraft in A_r, for each aircraft, we will predict
the possible location of the aircraft according to its last-known position, velocity,
and direction. And the real position of the aircraft is determined from m aircraft
near the possible location. Therefore, all aircraft will be divided into $|S|$ shares in
A_r, and the proportion of each share in the total is $\frac{m_i}{n}$ $(i = 1, 2, \ldots, |S|)$. Then
the probability of selecting a candidate in A_r is $P_i = \frac{m}{n}$. Hence, we have

$$H\left(p\right) = -\sum_{i=1}^{|S|} P_i log_2 P_i$$

$$= -\sum_{i=1}^{|S|} \frac{m_i}{n} log_2 \frac{m_i}{n} \qquad (2.14)$$

$$= log_2 n - \frac{1}{n}\sum_{i=1}^{|S|} m_i log_2 m_i$$

Assuming that m_i is a normal distribution with $\mu = \frac{n}{2}$ and $\sigma = \frac{n}{6}$, then we have

$$m_i = \left| \frac{1}{\sqrt{2\pi}\,\frac{n}{6}} \exp\left[-\frac{1}{2}\left[\frac{x - \frac{n}{2}}{\frac{n}{6}} \right]^2 \right] \right| \qquad (2.15)$$

2.3.4.2   Experimental Results

For the simple tracking method, we evaluated the location-privacy level relative to the silence period and airspace density. As shown in Fig. 2.18a,b, such the level improves with the silence period and airspace density, respectively. Specifically, the $hsep_{min}$ will be 6 km considering Class A airspace with an average of 900 km an hour.

For the correlation tracking method, we also analyzed the location-privacy level regarding the estimated number of times, N, and the anonymity set size $|S|$. Figure 2.18c demonstrates as N grows, this level begins to increase and then stabilizes. To evaluate this level, a proper N needs to be found. Figure 2.18d shows the level grows with $|S|$.

2.4   Conclusion

In this chapter, we first discussed the usability of modern cryptography in the ADS-B system. We have found that it is improper for existing cryptographic techniques to be straightforwardly applied to secure ADS-B. For example, the direct encryption of an entire message may violate the openness of the ADS-B system; in PKI-based cryptographic schemes, the verification and transmission of certificates will bring large computational and communication costs, which are not well-suited to resource-limited avionics and low-bandwidth data links. Therefore, in the following chapters, we will explore emerging cryptographic techniques to protect the security of the ADS-B system. Specifically, in Chap. 3, we suggest

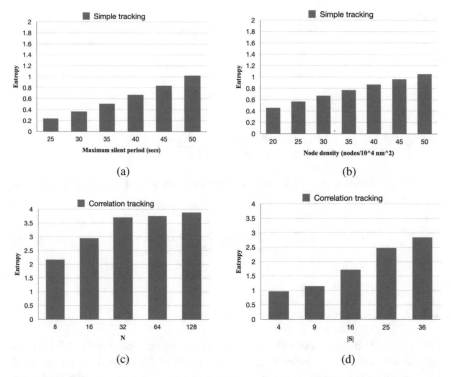

Fig. 2.18 Maximum location privacy of target. (**a**) Node Density $= 30/10^4$ nm^2. (**b**) $SP_{max} =$ 40 s. (**c**) $|S| = 16$. (**d**) $N = 32$

two broadcast authentication schemes in an identity-based setting that significantly reduce the burden of running PKI; in Chap. 4, we design a privacy-preserving aircraft location verification scheme that uses VHE to achieve the ciphertext-based similarity measurement; and in Chap. 5, we propose a complete ADS-B security solution, in which FPE is exploited to encrypt only the AA field, not the entire message, guaranteeing comparability with existing ADS-B protocols and maintaining the openness of the ADS-B system.

References

1. K. D. Wesson, T. E. Humphreys, and B. L. Evans, "Can cryptography secure next generation air traffic surveillance?" *IEEE Security and Privacy Magazine*, 2014.
2. M. Strohmeier, V. Lenders, and I. Martinovic, "On the security of the automatic dependent surveillance-broadcast protocol," *IEEE Communications Surveys & Tutorials*, vol. 17, no. 2, pp. 1066–1087, 2014.
3. D. R. Stinson, *Cryptography: theory and practice*. Chapman and Hall/CRC, 2005.

4. H. Cheng and Q. Ding, "Overview of the block cipher," in *2012 Second International Conference on Instrumentation, Measurement, Computer, Communication and Control*. IEEE, 2012, pp. 1628–1631.
5. L. Jiao, Y. Hao, and D. Feng, "Stream cipher designs: a review," *Science China Information Sciences*, vol. 63, no. 3, pp. 1–25, 2020.
6. C. Sanchez-Avila and R. Sanchez-Reillol, "The Rijndael block cipher (AES proposal): a comparison with DES," in *Proceedings IEEE 35th Annual 2001 International Carnahan Conference on Security Technology (Cat. No. 01CH37186)*. IEEE, 2001, pp. 229–234.
7. V. Rijmen and J. Daemen, "Advanced encryption standard," *Proceedings of Federal Information Processing Standards Publications, National Institute of Standards and Technology*, pp. 19–22, 2001.
8. D. J. Bernstein, "The poly1305-AES message-authentication code," in *International workshop on fast software encryption*. Springer, 2005, pp. 32–49.
9. J. M. Turner, "The keyed-hash message authentication code (HMAC)," *Federal Information Processing Standards Publication*, vol. 198, no. 1, 2008.
10. R. C. Merkle, "A digital signature based on a conventional encryption function," in *Conference on the theory and application of cryptographic techniques*. Springer, 1987, pp. 369–378.
11. R. Kaur and A. Kaur, "Digital signature," in *2012 International Conference on Computing Sciences*. IEEE, 2012, pp. 295–301.
12. R. Housley, W. Ford, W. Polk, D. Solo *et al.*, "Internet X. 509 public key infrastructure certificate and CRL profile," RFC 2459, January, Tech. Rep., 1999.
13. C. Adams and S. Lloyd, *Understanding public-key infrastructure: concepts, standards, and deployment considerations*. Sams Publishing, 1999.
14. J. Weise, "Public key infrastructure overview," *Sun BluePrints OnLine, August*, pp. 1–27, 2001.
15. A. Shamir, "Identity-based cryptosystems and signature schemes," in *Workshop on the theory and application of cryptographic techniques*. Springer, 1984, pp. 47–53.
16. D. Boneh and M. Franklin, "Identity-based encryption from the Weil pairing," in *Annual international cryptology conference*. Springer, 2001, pp. 213–229.
17. A. Sahai and B. Waters, "Fuzzy identity-based encryption," in *Annual international conference on the theory and applications of cryptographic techniques*. Springer, 2005, pp. 457–473.
18. J. Baek, J. Newmarch, R. Safavi-Naini, and W. Susilo, "A survey of identity-based cryptography," in *Proc. of Australian Unix Users Group Annual Conference*, 2004, pp. 95–102.
19. D. Boneh and X. Boyen, "Secure identity based encryption without random oracles," in *Annual International Cryptology Conference*. Springer, 2004, pp. 443–459.
20. L. Chen and J. Malone-Lee, "Improved identity-based signcryption," in *International workshop on public key cryptography*. Springer, 2005, pp. 362–379.
21. H. Li, Y. Dai, L. Tian, and H. Yang, "Identity-based authentication for cloud computing," in *IEEE international conference on cloud computing*. Springer, 2009, pp. 157–166.
22. R. Canetti, S. Halevi, and J. Katz, "A forward-secure public-key encryption scheme," in *International Conference on the Theory and Applications of Cryptographic Techniques*. Springer, 2003, pp. 255–271.
23. S. S. Chow, "Removing escrow from identity-based encryption," in *International workshop on public key cryptography*. Springer, 2009, pp. 256–276.
24. T. H. Yuen, W. Susilo, and Y. Mu, "How to construct identity-based signatures without the key escrow problem," *International Journal of Information Security*, vol. 9, no. 4, pp. 297–311, 2010.
25. D. Galindo, J. Herranz, and J. Villar, "Identity-based encryption with master key-dependent message security and leakage-resilience," in *European Symposium on Research in Computer Security*. Springer, 2012, pp. 627–642.
26. L. Purton, H. Abbass, and S. Alam, "Identification of ads-b system vulnerabilities and threats," in *Australian Transport Research Forum, Canberra*, 2010, pp. 1–16.
27. D. McCallie, J. Butts, and R. Mills, "Security analysis of the ads-b implementation in the next generation air transportation system," *International Journal of Critical Infrastructure Protection*, vol. 4, no. 2, pp. 78–87, 2011.

28. M. Strohmeier, M. Schäfer, V. Lenders, and I. Martinovic, "Realities and challenges of nextgen air traffic management: the case of ads-b," *IEEE Communications Magazine*, vol. 52, no. 5, pp. 111–118, 2014.

29. M. R. Manesh and N. Kaabouch, "Analysis of vulnerabilities, attacks, countermeasures and overall risk of the automatic dependent surveillance-broadcast (ADS-B) system," *International Journal of Critical Infrastructure Protection*, vol. 19, pp. 16–31, 2017.

30. A. Costin and A. Francillon, "Ghost in the air (traffic): On insecurity of ads-b protocol and practical attacks on ads-b devices," *Black Hat USA*, pp. 1–12, 2012.

31. K. Sampigethaya and R. Poovendran, "Security and privacy of future aircraft wireless communications with offboard systems," in *2011 Third International Conference on Communication Systems and Networks (COMSNETS 2011)*. IEEE, 2011, pp. 1–6.

32. K. Sampigethaya, S. Taylor, and R. Poovendran, "Flight privacy in the NextGen: Challenges and opportunities," in *2013 Integrated Communications, Navigation and Surveillance Conference (ICNS)*. IEEE, 2013, pp. 1–15.

33. R. Gauthier and R. Seker, "Addressing operator privacy in automatic dependent surveillance-broadcast (ADS-B)," in *Proceedings of the 51st Hawaii International Conference on System Sciences*, 2018.

34. R. Housley, *Public key infrastructure (PKI)*. John Wiley & Sons, Inc.: Hoboken, NJ, USA, 2004, vol. 3.

35. E. Cook, "ADS-B, friend or foe: ADS-B message authentication for NextGen aircraft," in *2015 IEEE 17th International Conference on High Performance Computing and Communications, 2015 IEEE 7th International Symposium on Cyberspace Safety and Security, and 2015 IEEE 12th International Conference on Embedded Software and Systems*. IEEE, 2015, pp. 1256–1261.

36. Z. Wu, T. Shang, and A. Guo, "Security issues in automatic dependent surveillance-broadcast (ADS-B): A survey," *IEEE Access*, vol. 8, pp. 122 147–122 167, 2020.

37. R. Bolczak, J. C. Gonda III, W. J. Saumsiegle, and R. A. Tornese, "Controller-pilot data link communications (CPDLC) build 1 value-added services," in *The 23rd Digital Avionics Systems Conference (IEEE Cat. No. 04CH37576)*, vol. 1. IEEE, 2004, pp. 2–D.

38. I. Sestorp and A. Lehto, "CPDLC in practice: a dissection of the controller pilot data link communication security," 2019.

39. H. Alanazi, B. B. Zaidan, A. A. Zaidan, H. A. Jalab, M. Shabbir, Y. Al-Nabhani *et al.*, "New comparative study between DES, 3DES and AES within nine factors," *arXiv preprint arXiv:1003.4085*, 2010.

40. G. Singh, "A study of encryption algorithms (RSA, DES, 3DES and AES) for information security," *International Journal of Computer Applications*, vol. 67, no. 19, 2013.

41. D. S. Hicok and D. Lee, "Application of ADS-B for airport surface surveillance," in *17th DASC. AIAA/IEEE/SAE. digital avionics systems conference. Proceedings (Cat. No. 98CH36267)*, vol. 2. IEEE, 1998, pp. F34–1.

42. K. Samuelson, E. Valovage, and D. Hall, "Enhanced ADS-B research," in *2006 IEEE Aerospace Conference*. IEEE, 2006, pp. 7–pp.

43. T. Kacem, D. Wijesekera, and P. Costa, "Integrity and authenticity of ADS-B broadcasts," in *2015 IEEE Aerospace Conference*. IEEE, 2015, pp. 1–8.

44. H. Yang, M. Yao, Z. Xu, and B. Liu, "LHCSAS: A lightweight and highly-compatible solution for ADS-B security," in *GLOBECOM 2017-2017 IEEE Global Communications Conference*. IEEE, 2017, pp. 1–7.

45. H. Yang, Q. Zhou, M. Yao, R. Lu, H. Li, and X. Zhang, "A practical and compatible cryptographic solution to ADS-B security," *IEEE Internet of Things Journal*, vol. 6, no. 2, pp. 3322–3334, 2018.

46. D. Johnson, A. Menezes, and S. Vanstone, "The elliptic curve digital signature algorithm (ECDSA)," *International journal of information security*, vol. 1, no. 1, pp. 36–63, 2001.

47. W.-J. Pan, Z.-L. Feng, and Y. Wang, "ADS-B data authentication based on ECC and X. 509 certificate," *Journal of Electronic Science and Technology*, vol. 10, no. 1, pp. 51–55, 2012.

48. R. V. Robinson, K. Sampigethaya, M. Li, S. Lintelman, R. Poovendran, and D. von Oheimb, "Secure network-enabled commercial airplane operations: It support infrastructure challenges," in *Proceedings of the First CEAS European Air and Space Conference Century Perspectives (CEAS)*, 2007.

49. Z. Feng, W. Pan, and Y. Wang, "A data authentication solution of ADS-B system based on X. 509 certificate," in *27th International Congress of the Aeronautical Sciences, ICAS*, 2010, pp. 1–6.

50. M. Bellare, T. Ristenpart, P. Rogaway, and T. Stegers, "Format-preserving encryption," in *International workshop on selected areas in cryptography*. Springer, 2009, pp. 295–312.

51. T. Spies, "Format preserving encryption," *Unpublished white paper, www. voltage. com Database and Network Journal (December 2008), Format preserving encryption: www. voltage. com*, 2008.

52. M. Bellare, P. Rogaway, and T. Spies, "The FFX mode of operation for format-preserving encryption," *NIST submission*, vol. 20, p. 19, 2010.

53. M. P. Integer, "Rational arithmetic cryptographic library (MIRACL)," 2018.

54. H. Zhou and G. Wornell, "Efficient homomorphic encryption on integer vectors and its applications," in *Proc. of ITA*, 2014, pp. 1–9.

55. S. Bogos, J. Gaspoz, and S. Vaudenay, "Cryptanalysis of a homomorphic encryption scheme," *Cryptography and Communications*, vol. 10, no. 1, pp. 27–39, 2018.

56. O. Regev, "On lattices, learning with errors, random linear codes, and cryptography," *Journal of the ACM*, vol. 56, no. 6, pp. 1–40, 2009.

57. A. Perrig, R. Canetti, J. D. Tygar, and D. Song, "The tesla broadcast authentication protocol," *Rsa Cryptobytes*, vol. 5, no. 2, pp. 2–13, 2002.

58. A. Perrig, D. Song, R. Canetti, J. Tygar, and B. Briscoe, "Timed efficient stream loss-tolerant authentication (TESLA): Multicast source authentication transform introduction," *Request For Comments*, vol. 4082, 2005.

59. M. Baugher and E. Carrara, "The use of timed efficient stream loss-tolerant authentication (TESLA) in the secure real-time transport protocol (SRTP)," RFC 4383, February, Tech. Rep., 2006.

60. V. Roca, A. Francillon, and S. Faurite, "Use of timed efficient stream loss-tolerant authentication (TESLA) in the asynchronous layered coding (ALC) and Nack-oriented reliable multicast (NORM) protocols," *IETF RFC 5776*, 2010.

61. D. Liu and P. Ning, "Multilevel μtesla: Broadcast authentication for distributed sensor networks," *ACM Transactions on Embedded Computing Systems (TECS)*, vol. 3, no. 4, pp. 800–836, 2004.

62. K. Sampigethaya, R. Poovendran, and C. S. Taylor, "Privacy of general aviation aircraft in the NextGen," in *2012 IEEE/AIAA 31st Digital Avionics Systems Conference (DASC)*. IEEE, 2012, pp. 7B5–1.

63. V. Herndon, "Itt,adsbexplained," Website, 2009, www.itt.com/adsb/adsb-explained.html.

64. K. Sampigethaya, M. Li, L. Huang, and R. Poovendran, "Amoeba: Robust location privacy scheme for VANET," *IEEE Journal on Selected Areas in communications*, vol. 25, no. 8, pp. 1569–1589, 2007.

Chapter 3
ADS-B Broadcast Authentication

3.1 Introduction

Since the ADS-B communication protocol does not have an inherent message authentication mechanism, ADS-B messages are vulnerable to message injection and modification attacks, so the authenticity and integrity of ADS-B broadcast messages cannot be guaranteed. Some broadcast authentication schemes have been proposed to ensure such authenticity and integrity, but they are not suitable for ADS-B communication scenarios where ADS-B transponders are resource-limited and ADS-B data links are low bandwidth. Specifically, most of the existing broadcast authentication schemes are certificate-based, and the verification and transmission of the certificate chain require a lot of computing and communication resources. Therefore, non-certificate-based broadcast authentication schemes are essential to achieve the authenticity and integrity of ADS-B messages with less verification time and lower communication bandwidth.

In this chapter, we propose two ADS-B broadcast authentication schemes, AuthBatch and AuthMR, both of which are based on IBS and no longer need PKI, reducing the costs of key management and certificate verification and transmission. Particularly, AuthBatch supports batch verification of multiple messages, and thus the average verification time for each message is shortened, which can be used for avionics with constrained resources; in AuthMR, the message is recovered from the signature, so the total length of the message and the signature is small, well-suited to low-bandwidth ADS-B data links.

The remainder of this chapter is organized as follows. In Sect. 3.2, we review related work. We then state the problem in Sect. 3.3. Next, we propose AuthBatch and AuthMR in Sects. 3.4 and 3.5, respectively. We conclude this chapter in Sect. 3.6.

© The Author(s), under exclusive license to Springer Nature Switzerland AG 2023
H. Yang et al., *Secure Automatic Dependent Surveillance-Broadcast Systems*,
Wireless Networks, https://doi.org/10.1007/978-3-031-07021-1_3

3.2   Related Work

The ADS-B protocol does not have a message authentication mechanism, so an attacker can inject fake aircraft into the ADS-B system. This spoofing attack seriously damages the authenticity and integrity of ADS-B messages and may threaten air traffic safety. Existing aerial surveillance technologies can thwart the threat of deception. For instance, by using SSR, a few bogus aircraft are filtered out, but the attacker can easily introduce many ghost aircraft onto the radar screen of ATCO, which vastly outstrips the processing capacity of SSR [1].

Cryptographic methods can also protect the authenticity and integrity of ADS-B messages. Encryption, for example, encodes broadcast messages into ciphertexts under the senders' keys. Unfortunately, the public is unable to decipher encrypted messages in the absence of keys. As a result, encryption clearly violates the ADS-B design aim of openness and cannot be used directly to prevent counterfeiting. Unlike encryption, the use of authentication does not change the original message; only authentication data is attached, such as a message authentication code (MAC) or digital signature. In this situation, the original message is available to all participants, no matter if the authentication data is checked, ensuring the openness of the ADS-B system.

Therefore, ADS-B message authentication (especially broadcast message authentication) is very important for air traffic safety. However, owing to the unique characteristics of ADS-B data links, this authentication has not been studied in depth in civil aviation. For example, Timed Efficient Stream Loss-tolerant Authentication (TESLA) is widely known as a broadcast authentication protocol with low computation and communication cost [2]. But TESLA has a large time delay and may not be well used for the authenticity of emergency broadcasts. Sampigethaya et al. proposed cryptographic methods to authenticate ADS-B messages [3, 4]. These methods require either the distribution of pre-shared secrets or the availability of PKI, posing scalability and cost issues for the ADS-B system. For instance, using a block cipher, Chen then suggested an authenticated encryption approach to preserve confidentiality and integrity for secure ADS-B data links [5]. Nevertheless, it is difficult to distribute secret keys in real-time since the key channel may not be well established. Moreover, the encryption of the entire message will be incompatible with open ADS-B systems. Through PKI and Elliptic Curve Data Signature Algorithm (ECDSA), Pan et al. then designed a message authentication solution for ADS-B communication [6]. Unfortunately, costly computation and communication are required to verify and transmit the certificate chain. As a result, this does not apply to ADS-B data links with low bandwidth. As a result, existing methods may not be appropriate for ADS-B data links with low bandwidth and poor connectivity.

As aforementioned, instead of individual signature verification, batch verification performs multiple message authentications once, and thus computational costs can be amortized. To guarantee efficient broadcast authentication for the ADS-B message, researchers have proposed some batch IBS methods [7–9]. Nevertheless, these methods sacrifice either security or efficiency. For instance, in [7], $n + 1$ costly pairing calculations were required to verify n signatures; in [8], attackers were capable of forging signatures passing batch verification; in [9], the security proof used an ideal random oracle that may not truly ensure security in the real world. In this work, we use Kurosawa-Heng IBS [10] to construct *AuthBatch*. The Kurosawa-Heng IBS has the following two advantages: First, its security is proven in the standard model; second, only one pairing is needed for signature verification. Note that what Kurosawa-Heng gave is indeed an identity-based identification. To convert it to the signature version, we use the Fiat-Shamir heuristic [11]. We further extend it to support batch verification.

In addition, the Identity-Based Signature with Message Recovery (IBS-MR) technique offers another promising scheme for ADS-B message broadcast authentication, which mainly includes two advantages: (1) in the IBS system, the user and the PKG, respectively, extract the user's public key and private key from the user's identity string, and thus the PKI is unnecessary and the cost of key management can be significantly reduced; and (2) messages can be recovered straightforwardly from the signature. No separate transmission is required, and the total length of the message and the corresponding signature decreases. IBS-MR may be suitable for low-bandwidth data links of ADS-B. Several practical IBS schemes [8, 12, 13] have been proposed since 1984 [14], but IBS-MR scheme has been proposed until 2005 [15]. However, the verification in [15] requires two expensive bilinear pairing computations. This thus inspires us to design efficient IBS-MR schemes.

3.3 Problem Statement

In this section, we state the problem, which includes the system model and threat model, design objectives, and preliminaries.

3.3.1 System Model and Threat Model

In our system model, aircraft first get GNSS navigation signals through equipped GNSS receivers. Aircraft then regularly broadcast ADS-B messages twice a second through built-in ADS-B transmitters. ADS-B communications between aircraft and aircraft or aircraft and ground stations are carried out via two subsystems, ADS-B-Out and ADS-B-in. There are two ADS-B data link standards, UAT [16] and 1090ES [17]. The 1090ES has become highly congested since ATC radar systems are currently using it. Therefore, this chapter focuses on the UAT-based ADS-B broadcast authentication. Besides, each aircraft possesses the 24-bit ICAO address, which is a globally coordinated permanent unique identifier. In the identity-based setting, we treat it as a unique identity.

Because of the broadcast nature of the ADS-B data link, ADS-B messages are susceptible to a variety of cyberattacks. The work in [18] divides them into two categories: passive attacks (e.g., eavesdropping) and active attacks (e.g., jamming and spoofing). In this chapter, we will concentrate on spoofing attacks—that is, attackers can insert fake aircraft or corrupt air traffic data. Besides, side-channel attacks may expose private key vulnerabilities by exploiting various key information leakages during ADS-B broadcast authentication [19]. Here we use key-evolution technology to mitigate the damage of private key exposure.

3.3.2   Design Objectives

The following design objectives are required to enable secure ADS-B broadcast authentication:

- *Integrity and authenticity.* The ADS-B messages transmitted are actually sent by legal aircraft or ground stations and should not be forged or falsified during communications.
- *Resilience to key leakage.* Even if compromising any previous private key of the broadcaster, the attacker cannot employ it now or in the future. The independence of private key evolution should be guaranteed.
- *Low cost.* Due to the limited bandwidth of ADS-B data links, the communication overhead should be low. Due to the constrained resources of avionics devices, the computational cost should be low.

3.3.3   Preliminaries

We present preliminaries including the ADS-B message format and bilinear map. First, Table 3.1 gives the used notations in our two authentication schemes.

Because we showed the ADS-B message format of 1090ES in Chap. 1, we will only look at the ADS-B message format of UAT in this chapter. The UAT includes two message types: the basic ADS-B message and the long one. Both contain the head data record (HDR), by which we can correlate distinct messages from an aircraft. In addition, a 5-bit field is contained in HDR to represent the type of payload in the message, and thus there are 32 payload types (labeled from 0 to 31) as shown in Fig. 3.1. Here, the Type 30 is used to fill in the signature data of our AuthBatch scheme.

We then give a brief introduction on the bilinear map [20]. Let q be a prime, and we denote two q-order additive groups by G_1 and G_2, respectively. We then denote one q-order multiplicative group by G_T. Let P and Q be the generators of G_1 and G_2, respectively. Let ϕ be a computable isomorphism from G_2 to G_1 with

Table 3.1 Notations in AuthBatch and AuthMR

Notation	Meaning
$[\cdot]^{k_1}$	The first k_1 bits
$[\cdot]_{k_2}$	The last k_2 bits
$\lvert \cdot \rvert$	Bit length
$\oplus$	XOR operator
$\parallel$	Concatenation operator
λ	Security parameter
G_1, G_2, G_T	Three q-order cyclic groups with $\lvert q \rvert = k_1 + k_2$
$\hat{e} : G_1 \times G_2 \to G_T$	Admissible bilinear map (pairing operator)
P, Q	Generators of G_1, G_2, respectively
g	$\hat{e}(P, Q)$
$H_1 : \{0, 1\}^* \to Z_q^*$	Hash function
$H_2 : G_T \to Z_q^*$	Hash function
$H_2 : \{0, 1\}^* \times G_T \to \mathbb{Z}_q^*$	Hash function
$F_1 : \{0, 1\}^{k_2} \to \{0, 1\}^{k_1}$	Hash function
$F_2 : \{0, 1\}^{k_1} \to \{0, 1\}^{k_2}$	Hash function
N, T	Time period number and length, respectively

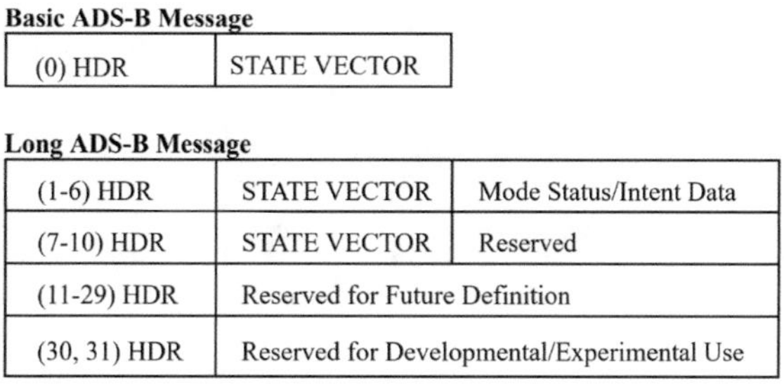

Fig. 3.1 UAT message format

Basic ADS-B Message

(0) HDR	STATE VECTOR

Long ADS-B Message

(1-6) HDR	STATE VECTOR	Mode Status/Intent Data
(7-10) HDR	STATE VECTOR	Reserved
(11-29) HDR	Reserved for Future Definition	
(30, 31) HDR	Reserved for Developmental/Experimental Use	

$\phi(Q) = P$. Then, we denote an admissible bilinear map by $\hat{e} : G_1 \times G_2 \to G_T$, which has three properties below:

- *Bilinear.* For $\forall X \in G_1$, $\forall Y \in G_2$, and $\forall m, n \in \mathbb{Z}_q^*$, we have $\hat{e}(mX, nY) = \hat{e}(X, Y)^{mn}$.
- *Computable.* $\hat{e}$ can be efficiently computed in a polynomial time.
- *Non-degenerate.* $\hat{e}(X, Y) \neq 1$.

Definition 3.1 We call a PPT algorithm $\mathcal{G}$ as *an asymmetric bilinear parameter generator* if taking a security level $\lambda \in \mathbb{Z}^+$ as input, $\mathcal{G}$ generates a seven-tuple $(q, P, Q, G_1, G_2, G_T, \hat{e})$, in which q is a λ-bit prime, G_1, G_2, and G_T are three q-order cyclic groups with $G_1 \neq G_2$, P, and Q are the generators of G_1 and G_2, respectively, and $\hat{e} : G_1 \times G_2 \to G_T$ is an admissible bilinear map.

Definition 3.2 The co-DH (co-Diffie-Hellman) problem is: Given $Q, aQ \in G_2$ and $P \in G_1$ as input for unknown $a \in \mathbb{Z}_q^*$, compute aP.

Definition 3.3 The co-DHI (co-Diffie-Hellman Inversion) problem is: Given $Q, aQ \in G_2$ and $P \in G_1$ as input for unknown $a \in \mathbb{Z}_q^*$, compute $a^{-1}P$. Note that from Mitsunari et al. [21], co-DH and co-DHI are equivalent in a polynomial time.

3.4   Batch Authentication

As shown in Fig. 3.2, a broadcaster B, whose identity is ID_B, uses its private key S_{ID_B} to generate an IBS signature $\theta = (r, S)$ of an ADS-B message m, and then broadcasts the message-signature pair (m, θ) via ADS-B-Out, which is allowed to

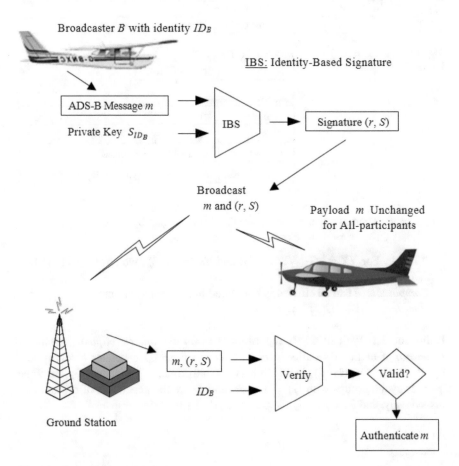

Fig. 3.2 Basic idea of AuthBatch

Fig. 3.3 Reserved field for signatures

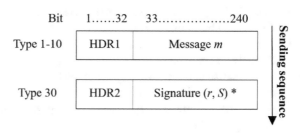

* Type 30 for accommodating Signature (r, S).

be verified by all receivers using ID_B. This message is sent in clear text so that all receivers can parse it according to specified message formats; the signature attached to this message provides capabilities of message verification. Those receivers who have requirements for the authenticity and integrity of messages may verify this signature; otherwise, they may not verify it. Regardless of whether this broadcast message is verified or not, all participants can see it, and thus our scheme is compatible with the original ADS-B protocols, keeping the openness of ADS-B systems. We will elaborate on the proposed scheme as follows.

3.4.1 Utilization of Reserved Field

Figure 3.3 illustrates how to use the reserved field of Type 30 to accommodate signatures. The ADS-B message of UAT has a data space of 272 bits, and considering that 32 bits have been used as HDR, only 240 bits can be used as payload [16]. We generate a new ADS-B message of such a reserved field to accommodate an IBS signature (r, S) that is then sent following an original ADS-B message of Type 1-10. Consequently, AuthBatch preserves the compatibility of ADS-B protocols by only attaching the new message of Type 30, but not altering the original message.

3.4.2 Details of AuthBatch

AuthBatch can be divided into five phases: *System Initialization*, *Broadcaster Registration*, *Signature Generation*, *Signature Verification*, and *Key Evolution*.

3.4.2.1 System Initialization

ATCO plays a role of PKG that initializes the AutoBatch system and publicly releases the system parameter as follows.

- *Step1*: Taking a security parameter λ, ATCO generates a 7-tuple $(q, P, Q, G_1, G_2, G_T, \hat{e})$ by running $\mathcal{G}(\lambda)$.
- *Step2*: ATCO randomly chooses $b \in \mathbb{Z}_q^*$ as the system master key that is kept privately and calculates $Q_{Pub} = bQ$ and $g = \hat{e}(P, Q)$.
- *Step3*: ATCO chooses two cryptographic hash functions $H_1 : \{0, 1\}^* \to \mathbb{Z}_q^*$ and $H_2 : \{0, 1\}^* \times G_T \to \mathbb{Z}_q^*$.
- *Step4*: ATCO publishes the public parameter as $\{q, P, Q, G_1, G_2, G_T, \hat{e}, Q_{Pub}, g, H_1, H_2\}$.

3.4.2.2   Broadcaster Registration

When a broadcaster B with the identity ID_B sends its registration request to the system, PKG works as follows.

- *Step1*: ATCO checks the validity of ID_B: if so, the process proceeds to the next step; otherwise, the process terminates.
- *Step2*: ATCO calculates the private key S_{ID_B} of ID_B as $S_{ID_B} = (b + H_1(ID_B))^{-1} P$.
- *Step3*: ATCO distributes S_{ID_B} to B through a secure channel. Note that the secure socket layer protocol can be used to setup a secure channel for the transmission of private keys [22].

3.4.2.3   Signature Generation

A broadcaster B uses its private key S_{ID_B} to generate a signature $\theta = (r, S)$ of a message $m \in \{0, 1\}^*$ as follows.

- *Step1*: B chooses randomly $k \in \mathbb{Z}_q^*$ and calculates $r = \hat{e}(P, Q_{Pub} + H_1(ID_B) Q)^k$.
- *Step2*: B sets $h = H_2(m, r)$ and calculates $S = kP + hS_{ID_B}$.
- *Step3*: B broadcasts the message-signature pair (m, θ) via ADS-B-Out. Note that B can also pre-calculate $\hat{e}(P, Q_{Pub} + H_1(ID_B) Q)$ to improve the efficiency of AuthBatch.

3.4.2.4   Signature Verification

Upon the received message-signature pair (m, θ) where $\theta = (r, S)$, each receiver performs the signature verification as follows.

- *Step1*: The receiver sets $h = H_2(m, r)$.
- *Step2*: The receiver checks whether $rg^h = \hat{e}(S, Q_{Pub} + H_1(ID_B) Q)$ holds. If yes, (r, S) is then a valid signature of m and the receiver accepts m; otherwise, the receiver discards m.

Receivers may simultaneously receive multiple message-signature pairs from different aircraft. Given l tuples $\{(ID_i, m_i, r_i, S_i)|i = 1, \cdots l\}$, each receiver carries out the batch signature verification as follows.

- *Step1*: The receiver sets $h_i = H_2(m_i, r_i)$ for $i = 1, \cdots l$.
- *Step2*: The receiver checks whether $\left(\prod_{i=1}^{l} r_i\right)\left(g^{\sum_{i=1}^{l} h_i}\right) = \hat{e}\left(\sum_{i=1}^{l} S_i, Q_{Pub}\right)$

  $\hat{e}\left(\sum_{i=1}^{l} H_1(ID_i) S_i, Q\right)$ holds. If yes, the batch signature verification succeeds and the receiver accepts all messages; otherwise, the receiver discards them. Note that this batch verification costs only l scalar multiplications and 2 pairings, thereby reducing verification time.

3.4.2.5 Key Evolution

To mitigate loss by private key leakage, we split the lifespan of AuthBatch into N time intervals (or periods) with the same interval length T, and the key evolution can be performed as follows.

- *Step1*: At the end of period i, B sends ATCO a request for the private key of $i + 1$.
- *Step2*: Upon receiving the request, ATCO will check identity validity. If B is a legitimate user, ATCO issues the private key of $i + 1$ as $S_{ID_B||i+1} = (b + H_1(ID_B||(d + (i + 1) * T)))^{-1} P$, where d is an initial period. Otherwise, the private key of period $i + 1$ is not generated and the private key of period i will be automatically revoked.
- *Step3*: ATCO returns $S_{ID_B||i+1}$ to B through a secure channel.

3.4.3 *Correctness and Security Analysis*

3.4.3.1 Correctness

The correctness of the signature verification and batch verification in AuthBatch is guaranteed based on the following two theorems.

Theorem 3.1 *It is correct for a single signature to be verified.*

Proof For simplification, we first denote $Q_{Pub} + H_1(ID_B) Q$ by Q_{ID_B} and then we have

$$\hat{e}(P, Q)^h$$

$$= \hat{e}\left(h(b + H_1(ID_B))^{-1} P, (b + H_1(ID_B)) Q\right)$$

$$= \hat{e}\left(h(b + H_1(ID_B))^{-1} P, Q_{Pub} + H_1(ID_B) Q\right)$$

$$= \hat{e}\left(h S_{ID_B}, Q_{ID_B}\right)$$

As a result, we have

$$\hat{e}\left(kP, Q_{ID_B}\right)\hat{e}\left(P, Q\right)^h = \hat{e}\left(kP, Q_{ID_B}\right)\hat{e}\left(hS_{ID_B}, Q_{ID_B}\right)$$
$$\Rightarrow \hat{e}\left(P, Q_{ID_B}\right)^k\hat{e}\left(P, Q\right)^h = \hat{e}\left(kP + hS_{ID_B}, Q_{ID_B}\right)$$
$$\Rightarrow rg^h = \hat{e}\left(S, Q_{ID_B}\right)$$
$$\Rightarrow rg^h = \hat{e}(S, Q_{Pub} + H_1(ID_B)Q)$$

Theorem 3.2 *It is correct for multiple signatures to be verified in a batch manner.*

Proof The correctness of batch verification can be justified as

$$\hat{e}\left(\sum_{i=1}^{l} S_i, Q_{Pub}\right)\hat{e}\left(\sum_{i=1}^{l} H_1(ID_i)S_i, Q\right)$$

$$= \hat{e}\left(\sum_{i=1}^{l} S_i, bQ\right)\hat{e}\left(\sum_{i=1}^{l} H_1(ID_i)S_i, Q\right)$$

$$= \hat{e}\left(\sum_{i=1}^{l} bS_i, Q\right)\hat{e}\left(\sum_{i=1}^{l} H_1(ID_i)S_i, Q\right)$$

$$= \hat{e}\left(\sum_{i=1}^{l}(b + H_1(ID_i))S_i, Q\right)$$

$$= \hat{e}\left(\sum_{i=1}^{l}(b + H_1(ID_i))(k_iP + h_iS_{ID_i}), Q\right)$$

$$= \hat{e}\left(\sum_{i=1}^{l}(b + H_1(ID_i))(k_iP), Q\right)\hat{e}\left(\sum_{i=1}^{l}(b + H_1(ID_i))(h_iS_{ID_i}), Q\right)$$

$$= \left(\prod_{i=1}^{l}\hat{e}((b + H_1(ID_i))(k_iP), Q)\right)\left(\prod_{i=1}^{l}\hat{e}((b + H_1(ID_i))(S_{ID_i}), h_iQ)\right)$$

$$= \left(\prod_{i=1}^{l}\hat{e}((k_iP), (b + H_1(ID_i))Q)\right)\left(\prod_{i=1}^{l}\hat{e}(P, h_iQ)\right)$$

$$= \left(\prod_{i=1}^{l}(\hat{e}(P, Q_{Pub} + H_1(ID_i)Q))^{k_i}\right)\left(\prod_{i=1}^{l}(\hat{e}(P, Q))^{h_i}\right)$$

$$= \left(\prod_{i=1}^{l} r_i\right)\left(g^{\sum_{i=1}^{l} h_i}\right)$$

Table 3.2 Functions of AuthBatch

Function	Pan et al. [6]	Chen [5]	AuthBatch
ADS-B authentication	✓	✓	✓
No certificate		✓	✓
No encryption	✓		✓
Standard model			✓
Batch verification			✓
Key evolution			✓

3.4.3.2 Authenticity and Integrity

We build AuthBatch security upon the Kurosawa-Heng IBS [10] which is unforgeable when facing chosen-message attacks without random oracles, and, therefore, the authenticity and integrity of ADS-B messages can be directly achieved.

3.4.3.3 Resilience of Key

In *Key Evolution*, the transmission of the next-period private key is via the secure channels between broadcasters and ATCO, which ensures adversaries who want to eavesdrop on communications cannot extract any useful information about the next-period private key. Even if adversaries compromise any previous private keys, they cannot infer any current or future private keys due to the intractability of discrete logarithm problems [22], and thus resilience for private key leakage can be guaranteed in AuthBatch.

In summary, we present a function comparison between AuthBatch with existing schemes as shown in Table 3.2 where the notation ✓ indicates that there is some function.

3.4.4 Performance Evaluation

We assess AuthBatch performance from computational time and communication bandwidth. In general, aircraft are resource-limited due to the small size of avionics, so performance evaluation focuses on aircraft.

3.4.4.1 Computational Time

We evaluate the execution time of each phase in AuthBatch through simulation based on the ARMv7 processor in the smartphone, which has an adjustable frequency from 200 MHz to 1440 MHz. We implement AuthBatch by using MIRACL

Table 3.3 Computational cost of AuthBatch (ms)

Frequency	T_{Init}	T_{Reg}	T_{Sig}	T_{Vrf}	T_{Evol}
200 MHz	654.2526	44.5630	89.1269	613.2895	48.1207
400 MHz	306.1514	21.5168	43.0336	291.6227	23.9273
800 MHz	149.2435	10.0965	20.1931	143.4164	11.4224
1000 MHz	119.8140	7.9800	15.9600	121.8339	8.2105
1440 MHz	84.9613	6.3856	12.7712	80.4850	6.6139

Table 3.4 Verification cost of AuthBatch

Scheme	1 signature	n signatures
Ours	$1\,Pa + 1\,M$	$2\,Pa + nM$
YCK [7]	$1\,Pa + 1\,M + 1\,H$	$(n+1)Pa + nM + nH$
ECDSA [24]	$4\,M$	$4\,nM$

cryptographic library (version 5.6.1) with the setting of AES-128 security and R-ate pairing on Barreto–Naehrig curve (embedding degree $k = 12$) with 1-2-4-12 tower of extensions [23]. We summarize the average running time (ms) of all phases, *System Initialization*, *Broadcaster Registration*, *Message Broadcast*, *Signature Verification*, and *Key Evolution*, which are denoted by T_{Init}, T_{Reg}, T_{Sig}, T_{Vrf}, and T_{Evol}, respectively. As illustrated in Table 3.3, even with the lowest frequency of 200 MHz, such time for each phase does not exceed 1 s, which demonstrates the high efficiency of AuthBatch suitable for avionics with constrained resources.

We use P_a, M, and H to, respectively, denote the operations of pairing, point multiplication in G_1, and map-to-point hash $H : \{0, 1\}^* \rightarrow G_1$. Note that the execution time of $H_1 : \{0, 1\}^* \rightarrow \mathbb{Z}_q^*$ and $H_2 : \{0, 1\}^* \times G_T \rightarrow \mathbb{Z}_q^*$ can be ignored. We then compare AuthBatch with YCK [7] and ECDSA schemes from verification cost, where ECDSA is a signature standard adopted in [6] and can be regarded as a comparison benchmark. Table 3.4 demonstrates operations used, respectively, for one-signature verification and n-signature batch verification. We observe that for n-signature verification, AuthBatch requires $2P_a + nM$ operations, while YCK and ECDSA require $(n + 1)P_a + nM + nH$ and $2nM$ operations, respectively. In particular, since ECDSA is certificate-based, extra operations are required to verify the public-key certificate (or certificate chain), so its overall verification cost is at least doubled.

We calculate the running time for P_a, M, and H based on a 1440 MHz ARM processor, which is 72.1901 ms, 6.3856 ms, and 8.2388 ms, respectively. Figure 3.4 further plots verification delay with the number of messages. We observe that the delay of using YCK is always the largest. When the number of messages is less than 8, ECDSA produces the smallest delay; but as the number of messages increases, the delay caused by AuthBatch is much less. We also find that the maximum number of signatures that can be verified by YCK, ECDSA, and AuthBatch are 3, 15, and 29, respectively, within a 300 ms time interval. Obviously, AuthBatch can verify the most signatures for a given time interval.

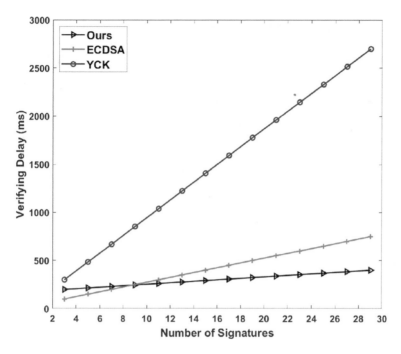

Fig. 3.4 Verification delay of AuthBatch

3.4.4.2 Communication Cost

We then compare AuthBatch with ECDSA on communication overheads that include a signature attached to a message but not the message itself. Note that an ECDSA signature is accompanied by a certificate (or a certificate chain). We take the 1024-bit RSA signature as the baseline security level. The signature length in AuthBatch is $|G_1| + |G_T|$ where the $|G_1| = 160$ and $|G_T| = 960$, which can be achieved by adopting 160-bit MNT curve [20]. Then the total signature length is 1120 bits (140 bytes). Although the ECDSA length is only 42 bytes, a certificate (or a certificate chain) is mandatory for practical usage. If the certificate adopts the type suggested by IEEE 1609.2 Standard [24], which has 125 bytes in length, the total transmission overhead of ECDSA is at least 167 (42 + 125) bytes. Figure 3.5 depicts the communication overhead with the number of messages. Obviously, the total length in AuthBatch is less than that in ECDSA. For this reason, it is more applicable for low-bandwidth ADS-B data link communications.

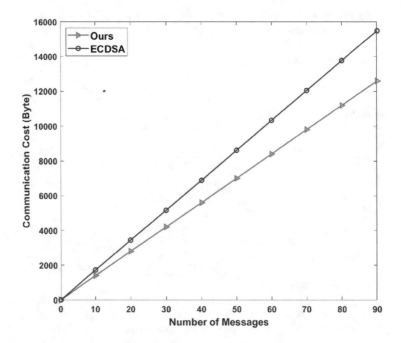

Fig. 3.5 Communication overhead of AuthBatch

3.5   Authentication with Message Recovery

We put forward a broadcast authentication scheme for ADS-B messages based on IBS-MR named AuthMR, which comprises the following five phases: *System Initialization, Broadcaster Registration, Message Signing and Broadcast, Verification and Recovery,* and *Key Evolution.*

3.5.1   Main Idea

The key thoughts of AuthMR are shown in Fig. 3.6: A broadcaster uses its private key to affix an IBS signature to each ADS-B transmission. This enables receivers to check the integrity of communications they have received, and any forgery or corruption is detectable. Conventional digital signature methods are computationally intensive and have a large transmission cost, making them unsuitable for ADS-B message authentication. Our AuthMR achieves lightweight computing (one pairing operation for verification). Besides the bandwidth savings associated with identity-based key distribution, recovering messages from signatures lowers

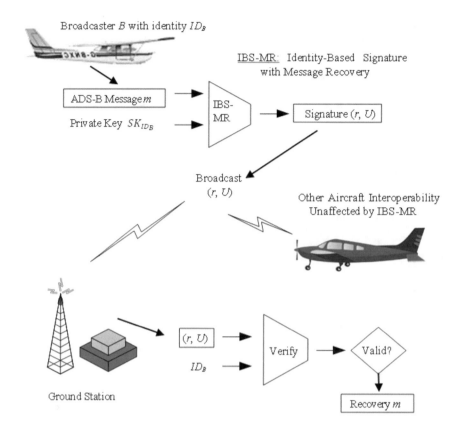

Fig. 3.6 Main idea of AuthMR

the overall length of messages and signatures, reducing the communication cost. Finally, AuthMR is resistant to key leaking because of private key evolution.

3.5.2 Details of AuthMR

3.5.2.1 System Initialization

ATCO can act as the PKG to set up all parameters as follows.

- *Step1:* Given a security parameter λ, run $\mathcal{G}(\lambda)$ to generate a 7-tuple $(q, P, Q, G_1, G_2, G_T, \hat{e})$.
- *Step2:* Choose a random number $s \in Z_q^*$, keeping it as the system master key secretly, and compute $Q_{Pub} = sQ$ and $g = \hat{e}(P, Q)$.

- **Step3:** Choose four secure cryptographic one-way hash functions H_1 : $\{0, 1\}^* \rightarrow Z_q^*$, H_2 : $G_T \rightarrow Z_q^*$, F_1 : $\{0, 1\}^{k_2} \rightarrow \{0, 1\}^{k_1}$, and F_2 : $\{0, 1\}^{k_1} \rightarrow \{0, 1\}^{k_2}$, where $k_1 + k_2 = |q|$.
- **Step4:** Publish the public parameters as

$$\{q, P, Q, G_1, G_2, G_T, \hat{e}, Q_{Pub}, g, H_1, H_2, F_1, F_2\}.$$

3.5.2.2   Broadcaster Registration

When broadcaster B wants to register her identity ID_B to the system, the PKG works in the following way:

- **Step1:** Check if ID_B is valid.
- **Step2:** Compute the private key SK_{ID_B} of ID_B as

$$SK_{ID_B} = (s + H_1(ID_B))^{-1} P.$$

- **Step3:** Send SK_{ID_B} to B through a secure channel.

Remark 1. It is assumed that a secure channel for transferring the private key can be accurately established using supplementary encryption tools like the wired or wireless SSL protocol.

3.5.2.3   Message Signing and Broadcast

Broadcaster B utilizes her private key SK_{ID_B} to sign a message $m \in \{0, 1\}^{k_2}$ and then broadcasts the signature (r, U):

- **Step1:** Choose a random $l \in Z_q^*$ and compute

$$v = g^l.$$

- **Step2:** To recover message from signature later, set

$$f = F_1(m)||(F_2(F_1(m)) \oplus m).$$

- **Step3:** Generate a signature (r, U) of m as

$$r = H_2(v) + f \ (mod \ q), \quad U = (1 + rl)SK_{ID_B}.$$

- **Step4:** Broadcast the signature (r, U) through ADS-B data link.

Remark 2. $|r + U| = |q| + |G_1|$ is the length of the signature, and the message m can be recovered from the signature, where $|m| = k_2$. Besides, a timestamp can be appended to the message m to avert replay attacks as common.

3.5.2.4 Verification and Recovery

Upon the receipt of (r, U), each recipient verifies the signature and recovers the message as follows:

- *Step1:* Compute

$$V = Q_{Pub}H_1 + (ID_B)Q.$$

- *Step2:* Compute

$$f = r - H_2\left(\hat{e}\,(U, V)^{r^{-1}}\,g^{-r^{-1}}\right) (mod\ q).$$

- *Step3:* Recover the message

$$m = [f]_{k_2} \oplus F_2([f]^{k_1}).$$

- *Step4:* Check if

$$[f]^{k_1} = F_1(m).$$

If the equation is correct, the signature will be accepted and the message m will be output.

3.5.2.5 Key Evolution

The private key can evolve over time in AuthMR to mitigate the harm due to the exposure of the private key. Specifically, the PKG divides the life cycle of AuthMR into N time periods, and for period $i \in \{1, \cdots, N\}$, the private key $SK_{ID_B||i}$ of broadcaster B with identity ID_B is calculated as

$$SK_{ID_B||i} = (s + H_1(ID_B||(d + i * T)))^{-1}P,$$

in which d is the initial time and T is the period length.

For broadcaster B, the PKG usually issues only one private key in each period through a secure channel between B and PKG. To reduce the high cost of retaining secure channels in a large-scale system, the PKG can issue n private keys once for the next periods, that is, the private keys are issued in the batch way. Note that an appropriate n can be a good trade-off between the issuing cost and secure level:

big n may reduce the communication cost but increase the key-exposure damage. Here we propose an adaptive key-evolution method. First, we denote the number of private keys B intends to apply at one time by n_B, and denote the maximum of n_B by m_B. To balance security and cost, the PKG may set m_B in an adaptive manner. Concretely, we list the procedure of private key evolution as follows:

- *Step1:* B first sets n_B at period i in terms of the specific security requirement.
- *Step2:* B then uploads n_B to PKG via a secure channel.
- *Step3:* Upon the recipient of n_B, PKG specifies n to be the smaller between m_B and n_B, calculates n private keys ω as

$$\omega = \{SK_{ID_B||(i+1)}, \cdots, SK_{ID_B||(i+n)}\},$$

  and returns ω to B through the secure channel.
- *Step4:* B secretly keeps ω for the next n-period use and further deletes previous private keys for security.

3.5.3   Correctness and Security Analysis

We present the correctness and security analysis for AuthMR in this section.

3.5.3.1   Correctness

Theorem 3.3 *The signature for the broadcast message is correct.*

Proof The correctness of AuthMR is demonstrated below.

$$\hat{e}(U, V)^{r^{-1}} g^{-r^{-1}}$$
$$= \hat{e}((1 + rl)SK_{ID_B}, Q_{Pub} + H_1(ID_B)Q)^{r^{-1}} g^{-r^{-1}}$$
$$= \hat{e}((1 + rl)(s + H_1(ID_B))^{-1}P, (s + H_1(ID_B))Q)^{r^{-1}} g^{-r^{-1}}$$
$$= \hat{e}((r^{-1} + l)P, Q)\hat{e}(P, Q)^{-r^{-1}}$$
$$= \hat{e}(P, Q)^{l}$$
$$= v$$

Hence, we have

$$r - H_2\left(\hat{e}(U, V)^{r^{-1}} g^{-r^{-1}}\right)(mod q)$$
$$= r - H_2(v)(mod q)$$
$$= f$$

In this way, the recipient can recover the message as $m = [f]_{k_2 \oplus F_2([f]^{k_1})}$. Since $f = F_1(m)||(F_2(F_1(m)) \oplus m)$, the recipient can easily check if $[f]^{k_1} = F_1(m)$ holds. Finally, the integrity of m is justified as

$$[f]_{k_2} \oplus F_2([f]^{k_1})$$
$$= F_1(m)||F_2(F_1(m)) \oplus m_{k_2} \oplus F_2(F_1(m))$$
$$= F_2(F_1(m)) \oplus m \oplus F_2(F_1(m))$$
$$= m$$

3.5.3.2   Unforgeability

Theorem 3.4 *Under a selected message attack, as long as the co-DHI issue is hard, the signature of the broadcast message cannot be forged in the random oracle model.*

Proof The proof is largely identical to that of the scheme in [15], and, therefore, we provide just a summary of the proof here. First, consider A as a probabilistic-polynomial-time adversary capable of achieving existential forgery on the signature in AuthMR with a non-ignorable advantage. Then there is an adversary B that solves the co-DHI issue with a non-ignorable advantage. Given $Q, sQ \in G_2$ where s is a random number in Z_q^*, B's goal is to compute $s^{-1}P$. To this end, B simulates a challenger and interacts with A as follows.

To begin with, B sets the system parameters $\{q, P, Q, G_1, G_2, G_T, \hat{e}, Q_{Pub}, g, H_1, H_2, F_1, F_2\}$, where $Q_{Pub} = sQ$, and H_1, H_2, F_1, and F_2 are random oracles controlled by B. And B guesses that A will select a certain ID in the challenge stage. Then, B runs A by giving A the system parameters. During the execution, B emulates A's oracles. Finally, A generates a signature (r, U) for the message m and ID with a non-negligible probability.

Then, B reloads and runs the game a second time. Through the use of the Forking Lemma [25], B acquires two distinct signatures for the same message m, (r, U) and (r', U'). Hence, B obtains $s^{-1}P = (r - r')^{-1}(U - U')$, which is the solution of co-DHI. However, there is no probabilistic polynomial-time method that solves the co-DHI issue with a non-negligible advantage. As a result of the above paradox, the unforgeability of broadcast messages is implied.

Table 3.5 Functions of AuthMR

Function	Chen [5]	Pan et al. [6]	AuthMR
ADS-B authentication	✓	✓	✓
No certificate	✓		✓
No encryption		✓	✓
Key evolution			✓
Message Recovery			✓

3.5.3.3   Independence

Theorem 3.5 *AuthMR achieves key independence for the evolution of private keys.*

Proof During the phase of Private Key Evolution, broadcaster B initially transmits n_B to PKG via a secure channel. Then, using n_B supplied by B, the PKG creates the set of private keys, ω, and transmits ω to B via the secure channel. Thus, even if an enemy A eavesdrops on the conversation between B and the PKG, it will be unable to learn anything about ω. Besides, admitting that the adversary A compromises the previous private key, it still cannot infer the current or future private key because the security of the private key is guaranteed by the hardness of the discrete logarithm problem. Therefore, AuthMR is able to achieve key independence for the evolution of private keys.

Finally, we give a comparison of security functions in Table 3.5.

3.5.4   Performance Evaluation

We assess the performance of AuthMR from computing cost and communication overhead. To begin, we summarize the performance of IBS systems. Then, we compare the transmission overhead with existing schemes that employ certificate-based signatures (ECDSA and X.509). Finally, we estimate the computational cost of AuthMR.

3.5.4.1   Performance Evaluation in General

We compare the overall performance of five IBS schemes in Table 3.6 in terms of the total transmission length ($|signature| + |message|$) and computational costs associated with signing and verification. The Pa, M, Exp, and H, respectively, represent a pairing operation, an elliptic curve scalar multiplication in G_1, an exponential operation in G_T, and a map-to-point hash function $H : \{0, 1\}* \rightarrow G_1$. Other operations are omitted here since their calculation costs are negligible compared with those listed above.

As shown in Table 3.6, the communication overhead needed by AuthMR and the scheme [15] is the same, less than the other three schemes. This is because both

Table 3.6 General performance evaluation of AuthMR

	Sign	Verify	Total length
Hess [13]	$1\,Exp + 2\,M$	$2\,Pa + 1\,Exp + 1\,m$	$|m| + |q| + |G_1|$
Yi [12]	$3\,M$	$2\,Pa + 2\,Exp + 1\,M$	$|m| + 2|G_1|$
Cha et al.[8]	$2\,m + 1\,M$	$2\,Pa + 2\,m + 1\,M$	$|m| + 2|G_1|$
ZSM [15]	$1\,Exp + 2\,M$	$2\,Pa + 1\,Exp + 1\,m$	$|q| + |G_1|$
AuthMR	$1\,Exp + 1\,M$	$1\,Pa + 1\,Exp + 1\,M$	$|q| + |G_1|$

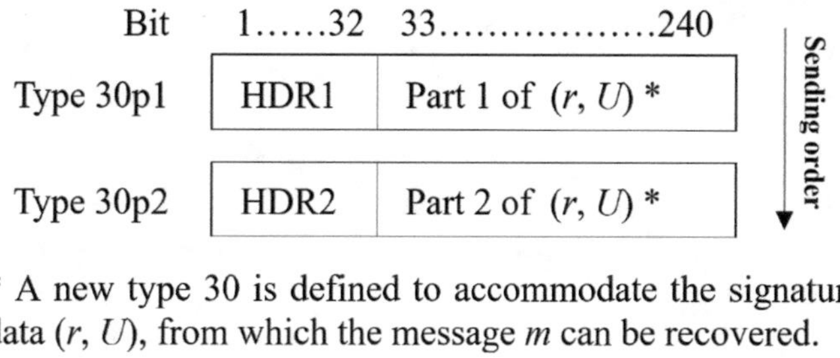

Bit 1……32 33………………240

Type 30p1 | HDR1 | Part 1 of (r, U) *

Type 30p2 | HDR2 | Part 2 of (r, U) *

Sending order

* A new type 30 is defined to accommodate the signature data (r, U), from which the message m can be recovered.

Fig. 3.7 UAT data composition

systems allow for the recovery of messages from signatures, making them well-suited to ADS-B low-bandwidth communication. Additionally, AuthMR reduces the computational cost of signing by 1M and validating by $1Pa$, making it more suitable for resource-constrained avionics equipment.

3.5.4.2 Communication Cost Comparison

Then we compare AuthMR and the method in [6] from transmission overhead. For convenience, we use the security level of 1024-bit RSA signature as our criteria. In [6], four UAT data frames are required to broadcast a signed message ($|m| = k_2$) because of the limited 240-bit data space available in UAT. As a result, the total length of $|signature| + |message|$ for each signed message is 960 bits. The total length of transmission in AuthMR is just 341 bits ($|q| + |G_1|$): If the 1024-bit RSA security level is needed, q should be a 170-bit prime, and G_1 should be a group whose elements are 171 bits using any curve families specified in [20]. Therefore, two UAT data frames are sufficient to accommodate a signed message (Fig. 3.7).

Figure 3.8 demonstrates the UAT frames needed for broadcasting a message with or without authentication. As shown, the number of frames in AuthMR is less than that in [6]. Note that the comparison excludes certificate transmission; otherwise, AuthMR has much less communication overhead owing to the inherent IBS setting.

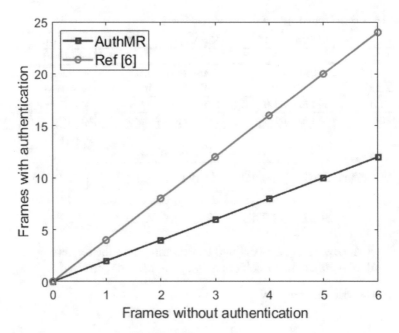

Fig. 3.8 Communication overhead of AuthMR

Table 3.7 Computational cost of AuthMR (ms)

	T_{Init}	T_{Reg}	T_{Sig}	T_{Vrf}
CPU = 200 MHz	609.6896	43.6453	65.5974	629.2988
CPU = 400 MHz	284.6346	20.7722	31.0010	294.0442
CPU = 800 MHz	139.1469	10.8369	15.5195	144.1214
CPU = 1000 MHz	111.8340	7.6892	12.8203	115.4903
CPU = 1440 MHz	78.5756	6.2214	9.1426	80.4287

3.5.4.3   Computational Time Estimation

We evaluate the running time for the major stages of AuthMR. Tests are performed using an ARMv7 processor with a frequency range from 200 MHz to 1440 MHz and 256M RAM. Then, we implement cryptographic operations using the MIRACL library (version 5.6.1), which employs R-rate pairing on the Barreto–Naehrig curve (embedding degree $k = 12$) with a 1-2-4-12 tower of extensions. AuthMR utilizes type-3 pairings, which are much more efficient than type-1 pairings. The computation time is described in Table 3.7, where T_{Init}, T_{Reg}, T_{Sig}, and T_{Vrf} indicate the average execution time (ms) for *System Initialization*, *Broadcaster Registration*, *Message Signing and Broadcast*, and *Verification and Recovery*, respectively. As seen, AuthMR is computationally efficient and well suited to avionics with constrained resources.

3.6 Conclusion

In this chapter, we have developed two ADS-B broadcast authentication schemes, AuthBatch and AuthMR, in which PKI and certificates are not required. In particular, AuthBatch supports batch verification of multiple messages, which is efficient for resource-constrained avionics; AuthMR recovers messages from signatures, cutting down the total length of messages and signatures, making them applicable for low-bandwidth ADS-B data links.

References

1. E. Valovage, "Enhanced ads-b research," in *2006 IEEE/AIAA 25TH Digital Avionics Systems Conference*. IEEE, 2006, pp. 1–7.
2. A. Perrig, R. Canetti, J. D. Tygar, and D. Song, "The tesla broadcast authentication protocol," *Rsa Cryptobytes*, vol. 5, no. 2, pp. 2–13, 2002.
3. K. Sampigethaya and R. Poovendran, "Privacy of future air traffic management broadcasts," in *2009 IEEE/AIAA 28th Digital Avionics Systems Conference*. IEEE, 2009, pp. 6–A.
4. K. Sampigethaya, R. Poovendran, S. Shetty, T. Davis, and C. Royalty, "Future e-enabled aircraft communications and security: The next 20 years and beyond," *Proceedings of the IEEE*, vol. 99, no. 11, pp. 2040–2055, 2011.
5. T.-C. Chen, "An authenticated encryption scheme for automatic dependent surveillance-broadcast data link," in *CSQRWC 2012*. IEEE, 2012, pp. 127–131.
6. W.-J. Pan, Z.-L. Feng, and Y. Wang, "ADS-B data authentication based on ECC and X. 509 certificate," *Journal of Electronic Science and Technology*, vol. 10, no. 1, pp. 51–55, 2012.
7. H. Yoon, J. H. Cheon, and Y. Kim, "Batch verifications with id-based signatures," in *International Conference on Information Security and Cryptology*. Springer, 2004, pp. 233–248.
8. J. C. Choon and J. H. Cheon, "An identity-based signature from gap Diffie-Hellman groups," in *International workshop on public key cryptography*. Springer, 2003, pp. 18–30.
9. P. S. Barreto, B. Libert, N. McCullagh, and J.-J. Quisquater, "Efficient and provably-secure identity-based signatures and signcryption from bilinear maps," in *International conference on the theory and application of cryptology and information security*. Springer, 2005, pp. 515–532.
10. K. Kurosawa and S.-H. Heng, "Identity-based identification without random oracles," in *International Conference on Computational Science and Its Applications*. Springer, 2005, pp. 603–613.
11. A. Fiat and A. Shamir, "How to prove yourself: Practical solutions to identification and signature problems," in *Conference on the theory and application of cryptographic techniques*. Springer, 1986, pp. 186–194.
12. X. Yi, "An identity-based signature scheme from the Weil pairing," *IEEE communications letters*, vol. 7, no. 2, pp. 76–78, 2003.
13. F. Hess, "Efficient identity based signature schemes based on pairings," in *International Workshop on Selected Areas in Cryptography*. Springer, 2002, pp. 310–324.
14. A. Shamir, "Identity-based cryptosystems and signature schemes," in *Workshop on the theory and application of cryptographic techniques*. Springer, 1984, pp. 47–53.
15. F. Zhang, W. Susilo, and Y. Mu, "Identity-based partial message recovery signatures (or how to shorten id-based signatures)," in *International Conference on Financial Cryptography and Data Security*. Springer, 2005, pp. 45–56.

16. R. DO, "282a, minimum operational performance standards for universal access transceiver (UAT) automatic dependent surveillance–broadcast (ADS-B)," 2004.

17. R. F. SC-186, *Minimum Operational Performance Standards for 1090 MHz Extended Squitter: Automatic Dependent Surveillance-Broadcast (ADS-B) and Traffic Information Services-Broadcast (TIS-B)*. RTCA, 2006.

18. D. McCallie, J. Butts, and R. Mills, "Security analysis of the ads-b implementation in the next generation air transportation system," *International Journal of Critical Infrastructure Protection*, vol. 4, no. 2, pp. 78–87, 2011.

19. F.-X. Standaert, T. G. Malkin, and M. Yung, "A unified framework for the analysis of side-channel key recovery attacks," in *Annual international conference on the theory and applications of cryptographic techniques*. Springer, 2009, pp. 443–461.

20. D. Boneh and X. Boyen, "Short signatures without random oracles," in *International conference on the theory and applications of cryptographic techniques*. Springer, 2004, pp. 56–73.

21. S. Mitsunari, R. Sakai, and M. Kasahara, "A new traitor tracing," *IEICE transactions on fundamentals of electronics, communications and computer sciences*, vol. 85, no. 2, pp. 481–484, 2002.

22. D. Boneh and M. Franklin, "Identity-based encryption from the Weil pairing," in *Annual international cryptology conference*. Springer, 2001, pp. 213–229.

23. E. Lee, H.-S. Lee, and C.-M. Park, "Efficient and generalized pairing computation on abelian varieties," *IEEE Transactions on Information Theory*, vol. 55, no. 4, pp. 1793–1803, 2009.

24. D. Jiang and L. Delgrossi, "IEEE 802.11 p: Towards an international standard for wireless access in vehicular environments," in *VTC Spring 2008-IEEE Vehicular Technology Conference*. IEEE, 2008, pp. 2036–2040.

25. D. Pointcheval and J. Stern, "Security arguments for digital signatures and blind signatures," *Journal of cryptology*, vol. 13, no. 3, pp. 361–396, 2000.

Chapter 4
Aircraft Location Verification

4.1 Introduction

In modern air traffic management, aircraft positioning is an important capacity to avoid flight collision and missing (e.g., MH370 disappearance [1]), where the locations of aircraft are constantly monitored to tell pilots and ATCOs. As the cornerstone of NextGen [2], ADS-B forces aircraft to broadcast periodically their positions gained from GNSS to improve situational awareness and extend surveillance range [3, 4]. However, because of the lack of security considerations when designing ADS-B, it is vulnerable to location spoofing attacks [5]. As an example, with cheap software-defined radios (SDRs), an attacker can inject fake locations to simulate trajectories of ghost airplanes into ATC, misleading collision-avoidance systems [6, 7]. Thus, certifying location claims in air traffic management is of extraordinary importance.

Cryptographic message authentication can confirm claimed locations [4, 8], but cryptographic improvements may face legislative and technical complexities for cross-border use [9]. Traditional wireless broadcast networks may use non-cryptographic techniques to ensure location legality [10–12]. Nevertheless, the ADS-B surveillance network has different characteristics from traditional wireless broadcast networks, such as fewer multi-path effects, out-of-sight scenes and long-distance transmission [13], needing specially designed aircraft location verification (ALV) technologies. Multilateration (MLAT) provides redundancy to double-check location claims by exploiting the time difference of arrival (TDOA) to pinpoint aircraft positions. Compared with radars, it improves verification accuracy [14]; but, MLAT requires at least four ADS-B sensors and performs worse under harsh geometric conditions (such as high mountains), which increases the difficulty of its widespread deployment in ADS-B surveillance networks. To overcome MLAT drawbacks, Strohmeier et al. used the grid-based k-nearest neighbor algorithm (kNN) for location confirmation in the predefined airspace, which is split into many small squares [15]. This method only requires three sensors and, in the meantime,

H. Yang et al., *Secure Automatic Dependent Surveillance-Broadcast Systems*, Wireless Networks, https://doi.org/10.1007/978-3-031-07021-1_4

reduces the influence of geometric dilution, so it is suitable for large-scale ADS-B deployment. The grid size should be small enough to improve verification accuracy. For a typical square of $75 \times 75 \text{ m}^2$, the surveillance plane of $33,000 \text{ km}^2$ will consist of six million grid squares, and then location verification needs to calculate six million Euclidean distances between the target aircraft and each square. The huge computational load makes it impossible to timely detect false position claims, which carries high flight safety risks.

For efficient and accurate validation of aircraft positional claims, we may outsource the kNN task and massive positional data to a strong third party, e.g., a public or private cloud [16–18]. However, the public cloud is not entirely trustworthy and thus may incur location-privacy issues, as aircraft locations can be used to infer future commercial and geopolitical transactions [19]. On the other hand, ATC and private aviation clouds [17, 18] may not completely trust each other [20]. In case of entrusting such positional information to a private aviation cloud, ATC may relinquish the physical control of sensitive positional information, jeopardizing the location privacy of aircraft. To solve this privacy issue, encrypting location data prior to outsourcing is a straightforward thought [21–24]. Nonetheless, conventional ciphers such as AES are incapable of validating encrypted locations owing to lacking ciphertext-manipulating capabilities. HE can support encrypted computation, but typical HEs, such as Paillier HE [25] or fully HE [26], exhibit considerable inefficiency. Recently, Zhou et al. presented an efficient VHE [27], which provides a choice to the large-scale location validation in the encrypted domain. Here, we will utilize VHE to design a privacy-preserving and efficient ALV scheme. This is not a simple work to use VHE to secure ALV and faces twofold challenges. First, the accuracy and efficiency of ALV are heavily dependent on the grid plane design, while previous designs did not pay attention to aircraft-altitude-change influences, which directly contributes to TDOA calculations [15, 28]. Second, it is challenging for the cloud to perform a non-interactive and efficient measure for encrypted Euclidean distances between the target aircraft and all grid squares in the surveillance area under no decryption [29].

In this chapter, we propose a privacy-preserving ALV scheme named PPALV simultaneously ensuring accuracy, efficiency, and privacy for the validation of aircraft positional claims. Specifically, our contributions are as follows.

– We utilize grid-based kNN techniques to develop PPALV for the validation of ADS-B location claims. In particular, we design flexible grid planes for kNN regression over the ATC surveillance area considering aircraft-altitude-change influences. Based on VHE, we also design an efficient and non-interactive encrypted-similarity-evaluation approach, which can be used to find k nearest encrypted squares.
– We provide a fast aircraft legitimacy recognition method, which only verifies the claimed position and does not estimate the real one. Therefore, this does not verify all the squares across the monitoring plane but only involves a few squares in a small circle in the entire plane, reducing the verification time.

– According to the security analysis, PPALV guarantees the privacy of the air-craft position and grid parameters. Performance evaluation shows that PPALV achieves almost the same accuracy as its corresponding plaintext version on OpenSky [30] (a sensor network used to collect ADS-B flight data). In addition, PPALV has high operating efficiency. Even with six million grids, it takes less than 200 ms, including calculation and memory transmission time, to evaluate the encryption distance.

The remainder of this chapter is structured in the following way: In Sect. 4.2, we survey related work. We state the preliminaries in Sect. 4.3. Then we present the problem in Sect. 4.4. In Sect. 4.5, we elaborate on PPALV, followed by an extended discussion in Sect. 4.6. Next, security analysis and experimental evaluation are given in Sects. 4.7 and 4.8, respectively. Finally, we conclude this chapter in Sect. 4.9.

4.2 Related Work

The basic idea of ALV is to double-determine the validity of claimed locations offered by both aircraft and other ADS-B participants. In doing so, the other participant also provides a method for determining the exact position of an aircraft, thus enabling redundancy to verify positional claims given by aircraft. Although numerous studies have been conducted on the secure location validation problem in conventional wireless broadcast networks [10–12], aircraft surveillance networks exhibit some unique characteristics such as large distance, outdoor line-of-sight, and few multi-path effects. For example, although TDOA systems have limitations in interior settings (owing to the multi-path effect), they thrive in long-distance outdoor line-of-sight conditions ideal for ATC. As a result, many aircraft location validation methods based on TDOA have been suggested, most notably MLAT and kNN [14, 15, 28].

MLAT is a self-contained method of localization that makes use of TDOA signals collected at many ground sensors [14], increasing localization accuracy compared with prior radar surveillance. MLAT, however, requires very precise synchronization and is strictly limited by the receiving point geometry. As a result, large-scale deployment of MLAT is prohibitively expensive. In comparison, the grid-based kNN method demonstrates successful localization even with low-cost hardware (e.g., SBS-3) and arbitrary receiver placement, which is advantageous for widespread implementation [15, 28]. Regrettably, high accuracy requires an unusually large number of grid squares. Due to the huge computational cost, the effectiveness of localization cannot be guaranteed, which will pose a grave hazard to flight safety.

With the proliferation of cloud computing, the job of ALV may be sent to the cloud to make use of plentiful computational resources [31]. Unfortunately, the cloud is often considered to be unreliable, making privacy protection for location-based services (LBS) an important issue. Hence, researchers have suggested many privacy-preserving LBS techniques, including [32, 33], and [34]. However, the majority of them need intrusive location data via anonymity [32, 35], or differential privacy [33]. For example, Gruteser et al. [32] proposed a geographical and temporal cloaking technique for LBS anonymity. Additionally, Miguel et al. [33] described the geo-indistinguishability of differential privacy for LBS and provided a method for perturbing locations taken from a planar Laplace distribution, but the disruption may not fully provide location privacy [36]. Also, the location assessment based on disrupted data may be insufficiently precise. Besides, some methods require a trusted third party [37], which most likely results in a performance bottleneck. Additionally, a distributed protocol based on the peer-to-peer architecture has been suggested to guarantee the privacy of LBS systems [38]. Regrettably, multiple rounds of interaction would incur significant communication costs.

Therefore, allowing cloud-based ALV while maintaining accuracy and efficiency is a key issue that must be resolved.

4.3 Preliminaries

In this section, we give preliminaries, including the OpenSky sensor network and two commonly used TDOA methods, MLAT and kNN.

4.3.1 OpenSky Sensor Network

OpenSky,[1] as illustrated in Fig. 4.1, is a non-profit research project jointly sponsored by Switzerland, Germany, and the UK in 2012. The OpenSky Network Association was established in 2015 to ensure the continuous development of OpenSky towards a fully open ATC sensor network covering the world.

OpenSky aims to improve the safety, reliability, and efficiency of airspace use by providing the public with open access to real-world ATC. It consists of many sensors connected to the Internet, owned by volunteers, industry supporters, and academic/government organizations [39]. A large historical database is used to collect raw data for researchers in different fields to analyze. The main technologies behind OpenSky are ADS-B and Mode S. These technologies provide detailed and real-time aircraft information through a publicly accessible 1090 MHz radio channel.

OpenSky started with eight sensors in Switzerland and then experienced an upsurge to extend to over 3000 registered receivers worldwide. Although the network initially concentrated only on ADS-B, in March 2017, it expanded its data range to include full S-mode down links. The data set currently contains over 22

---

[1] https://opensky-network.org/.

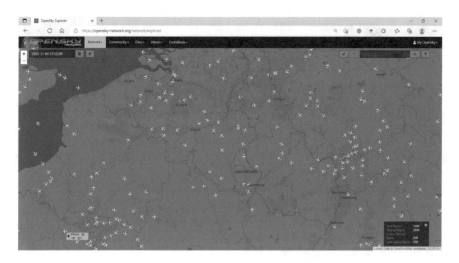

Fig. 4.1 OpenSky sensor network

trillion Mode S responses. Before the COVID pandemic, over 20 billion messages were received every single day. However, starting in March 2020, the number of daily messages has dropped to about 30% of the previous level, reflecting the reduction in global air travel because of the COVID-19 pandemic [40].

4.3.2 Time Difference of Arrival

One significant disadvantage of ADS-B is that incorrect location reports caused by avionics faults or spoofing attacks may interrupt ATC operations [5]. To deal with this, the technologies of TDOA-based aircraft location verification have been proposed. Figure 4.2 gives a framework of such technologies. In specific, an aircraft first acquires its locations from the global navigation satellite system. Then, the aircraft broadcasts ADS-B messages, including claimed locations, through ADS-B-Out. The adjacent aircraft and ground stations equipped with ADS-B receivers can obtain these messages through ADS-B-In. Subsequently, multiple ground stations measure the time of receiving the same message from the aircraft. They then transmit this data to the central processing station (CPS), which calculates the aircraft's real locations using TDOA. Finally, CPS sends the generated position reports (including claimed locations and calculated locations) to the ATC to perform ALV. In the following, two main TDOA-based ALV methods, MLAT and kNN, are given.

MLAT MLAT is a core technology of the air traffic management system for target surveillance of the airport scene, airway, and terminal area, as shown in Fig. 4.3.

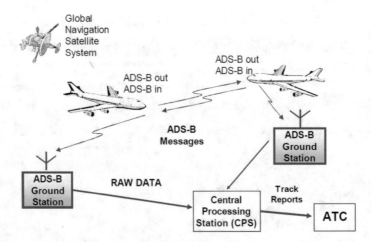

Fig. 4.2 Framework of using TDOA in ADS-B

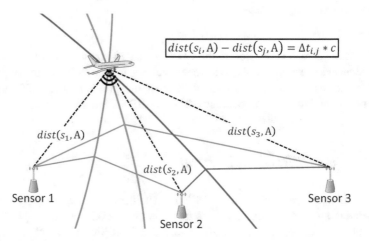

$$dist(s_i, A) - dist(s_j, A) = \Delta t_{i,j} * c$$

Fig. 4.3 Implementation of MLAT using TDOA

It is based on the TDOA: An aircraft broadcasts an ADS-B message, which is quickly received by many ADS-B sensors on the ground; the arriving time values at the sensors are accumulated and then computed as the TDOA values. Finally, hyperboloids based on TDOA are generated, and the cross point of hyperboloids is gained as the estimated position of the aircraft. However, MLAT also has some shortcomings when being used in ADS-B systems. First, MLAT needs at least four ADS-B sensors to receive aircraft data together. Second, its accuracy is highly related to geometrical dilution, which potentially leads to the inaccuracy of ALV under the conditions of complex terrain. Therefore, MLAT is improper for large-scale deployment.

kNN To overcome the disadvantages of MLAT, there is an alternative technology called kNN for estimating locations, which may rely on only three ADS-B sensors [41]. This technology leverages grid training regression, which constructs a rectangular grid plane and then break it down into numerous grid cells. Concretely, assuming that n fixed ADS-B sensors in the region, then for every grid cell, there exists a n-dimensional TDOA vector (also called fingerprint). When an aircraft flies into this area, it also produces a TDOA fingerprint for such n sensors. Then, the fingerprint similarity between the aircraft and each grid cell is evaluated by measuring Euclidean distance as

$$dist_{R,F} = \sqrt{\sum_{i=1}^{n} (R_i - F_i)^2} \tag{4.1}$$

where R_i and F_i, $i = 1, \cdots, n$, are the i-th element of n-dimensional fingerprints of the aircraft and grid cell, respectively, and $dist_{R,F}$ is the Euclidean distance. As shown in Fig. 4.4, the kNN needs to find k cells with the k smallest distances, and then the average of them is the estimated position of the aircraft. Further, if the gap between estimated and claimed positions is larger than the pre-assigned threshold, an alarm is issued to imply that the aircraft may potentially be a ghost aircraft.

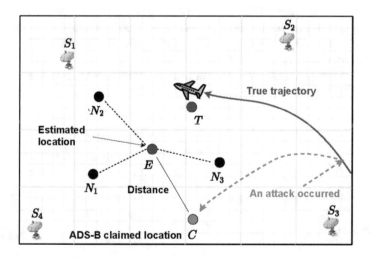

Fig. 4.4 Position estimation through kNN

4.4 Problem Statement

This section provides the problem statement that includes the system model, threat model, and design objectives.

4.4.1 System Model

Figure 4.5 demonstrates the system model consisting of five participants as follows.

- *Aircraft.* An aircraft equipped with an ADS-B transponder transmits its positional claims to the ground station within the signal propagation range.
- *Ground station.* Ground stations coupled with ADS-B receivers obtain ADS-B messages containing claimed positions and record their arrival time at the stations. The claimed locations and arrival time are then transmitted to CPS.
- *Central processing station (CPS).* CPS first calculates TDOA values for varying arrival times of the same ADS-B message to multiple ADS-B stations, and then encrypts these values and sends them to the cloud. On the other hand, CPS receives encrypted kNN results from the cloud, decrypts them, calculates the estimated position, and finally sends it to ATC.
- *Air traffic controller (ATCO).* ATC determines the legality of location claims after receiving claimed locations and estimated locations from CPS. If the answer is affirmative, ATC shows it on monitoring panels. Besides, ATC initializes the whole ALV system.
- *Cloud server (CS).* CS executes grid-based kNN tasks over large encrypted data from CPS and subsequently returns results to CPS.

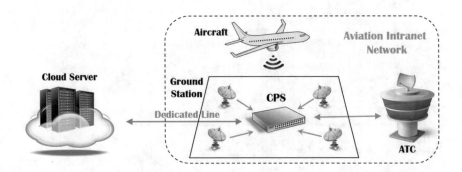

Fig. 4.5 System model of PPALV

4.4.2   Threat Model

The high-speed aviation intranet connects the aircraft, GS, CPS, and ATC, and we may consider them all to be part of the same trust domain [16–18]. They communicate securely and efficiently, and data may be sent in plain text without encryption. We often consider the cloud semi-honest and is not part of the same trust domain. As a result, the cloud is interested in eavesdropping on aircraft location privacy for commercial and other reasons. In particular, if we immediately send the sensitive data to the cloud, the cloud may determine the aircraft's claimed locations and estimated locations during verification processing. Even during the training phase, the cloud may learn the fingerprints and positions of grids in the surveillance region that are not accessible to the public aviation network. It is also worth mentioning that the connection between the aviation intranet and the cloud server is carried out through CPS, and their communication may take place via high-speed dedicated lines, as shown in Fig. 4.5, ensuring the transmission to be authentic, integral, and efficient.

To be summarized, our threat model focuses on location-privacy concerns caused by the cloud, which is honest-but-curious and motivated to spy on the location privacy of the aircraft for commercial or other reasons. Specifically, although the cloud faithfully performs the calculation of aircraft position verification according to the prescribed algorithm, it may not change verification results. However, the cloud has a strong curiosity, hoping to understand the location of the aircraft by outsourcing data. For this reason, the cloud launches a privacy breach attack.

4.4.3   Design Objectives

The overall goal is to conduct accurate and effective aircraft position verification while maintaining the confidentiality of the aerial surveillance network. This requires the following design objectives to be achieved.

- *Accuracy*. It should be ensured that the high accuracy required for aircraft position verification in the ciphertext domain is maintained.
- *Security*. Two types of data should have their confidentiality protected: 1) Position data for aircraft, including claimed positions and TDOA fingerprints; 2) Grid data, which includes grid coordinates and fingerprints. Additionally, the secure distance metric should be fulfilled.
- *Performance*. The verification time needs to be limited to ensuring that the ATC can effectively track the aircraft.

4.5 Proposed Scheme

The grid plane is first designed, considering the effect of aircraft altitude change.
We then develop a similarity measurement method for encrypted fingerprints. On
this basis, we propose a privacy-preserving ALV solution.

4.5.1 Grid Design with Altitude Change

Generally speaking, the cruising altitude of the aircraft is 11,000 m. However, with
the rapid increase in jet traffic, aircraft capacity at a height of 11,000 m has become
crowded. As a result, the cruising altitude today is between 9000 and 13,000 m—
the altitude range favored by airlines for fuel efficiency. Besides, to improve flight
safety, the monitoring scope not only covers air route cruises but also covers airports,
runways, and terminals, as illustrated in Fig. 4.6. Consequently, the effect of altitude
variation on ALV requires additional investigation.

As demonstrated in Fig. 4.7, Sensors 1 and 2 are situated at ground points P_1
and P_2, respectively, while the airplane is located at grid plane point P_3. It is
important to note that sensors may be installed at different heights above the ground.
However, compared with aircraft cruising altitude, the height of the sensors is almost
imperceptible. Aircraft routinely voyage at cruising altitudes of above 11,000 m,
whereas sensors are generally deployed at altitudes of 0 to 500 m.

Between the sensor and the airplane, the horizontal and vertical distances are d_h
and d_v, respectively, and such actual distance should be d_a. The aircraft altitude shift
in the vertical direction is Δ_d, and the actual distance measured is d_m. $|\Delta d|$ is much
smaller than $2d_v$ in this case since $|\Delta d|$ is often less than 2000 m and $2d_v$ is more
than 18,000 m. As a result, we have

$$d_a = \sqrt{d_h^2 + d_v^2}, \; d_m = \sqrt{d_h^2 + (d_v + \Delta d)^2} \tag{4.2}$$

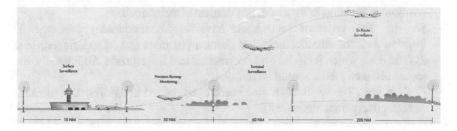

Fig. 4.6 Aircraft surveillance with altitudes

Fig. 4.7 Influence of altitude change

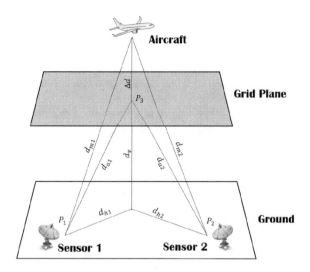

Here TDOA is utilized to calculate the position of the aircraft, it is essential to determine the distance between the aircraft and two distinct sensors, say Sensor 1 and Sensor 2. The horizontal distances d_{h1} and d_{h2} between the aircraft and these two sensors are different, but the vertical distances are equal ($d_v = d_{v1} = d_{v2}$) since they are referring to the same grid plane. As a result, we have

$$d_{a1} = \sqrt{d_{h1}^2 + d_v^2}, \; d_{m1} = \sqrt{d_{h1}^2 + (d_v + \Delta d)^2}, \tag{4.3}$$

$$d_{a2} = \sqrt{d_{h2}^2 + d_v^2}, \; d_{m2} = \sqrt{d_{h2}^2 + (d_v + \Delta d)^2} \tag{4.4}$$

Let $a = d_{h1}^2 + d_v^2$ and $b = d_{h2}^2 + d_v^2$. The initial distance difference between two sensors and the aircraft is calculated as

$$d_1 = \sqrt{a} - \sqrt{b} \tag{4.5}$$

and taking altitude change into account, the difference in measured distances is calculated as

$$d_2 = \sqrt{a + e} - \sqrt{b + e} \tag{4.6}$$

where $e = \Delta d(\Delta d + 2d_v) \approx 2d_v \Delta d$ since $|\Delta d|$ is much less than $2d_v$.

To evaluate the change of TDOA ($\Delta t = (d_1 - d_2)/c$, where c is the light speed), the value of $d = (d_1 - d_2)$ is also calculated as

$$d = d_1 - d_2 = (\sqrt{a} - \sqrt{a + e}) - (\sqrt{b} - \sqrt{b + e}) \tag{4.7}$$

To obtain d_{max}, the maximum of d for altitude change, we denote the maximum of d_1 and the minimum of d_2 by $max(d_1)$ and $min(d_2)$, respectively. Since both can be acquired at the same time, we have

$$d_{max} = max(d_1) - min(d_2). \tag{4.8}$$

It is easy to see that d_i is a function of the horizontal distance d_{hi} between the aircraft and Sensor i. Consequently, we have

$$d_i = \sqrt{x_i} - \sqrt{x_i + e} \tag{4.9}$$

where $x_i = d_{hi}^2 + d_v^2$, $i = 1, 2$. Without loss of generality, the function relationship can be expressed as

$$y = \sqrt{x} - \sqrt{x + e} \tag{4.10}$$

where $x = d_h^2 + d_v^2$.

We show the function relationship in Fig. 4.8, where $d_v = 11,000$ m, $|\Delta d| = 2000$ m and $0 < d_h < 100,000$ m, with respect to the detection range of typical ADS-B sensors being 100 km. It is worth noting that y ranges from the maximum of 950.1 m to the minimum of 99.4 m. As a result, we get $max(d_1) = 950.1$ m and $min(d_2) = 99.4$ m. Furthermore, we get $d_{max} = max(d_1) - min(d_2) = 850.7$ m, which means that d_{max} may be reached if one sensor is approximately under the

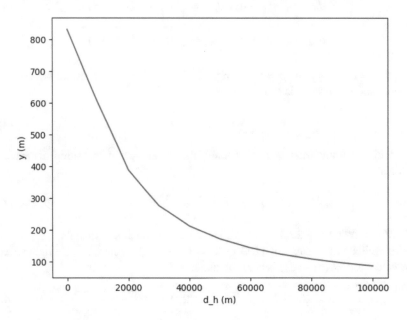

Fig. 4.8 Relationship between y and d_h

Table 4.1 Influence of altitude change

Altitude change [m]	Horizontal error [m]	
	Mean	Median
500	237.25	174.96
1000	341.61	251.32
1500	449.62	341.3
2000	609.93	456.68

aircraft and the other is from afar. Then, the equivalent time change for the light speed c is 2.84 us, which does not affect the calculation of TDOA.

Finally, we conduct tests to show the effect of altitude variation on ALV accuracy. You can refer to the experiment setting in Sect. 4.8.1. The distances between actual and estimated locations have a positive association with $|\Delta d|$, as seen in Table 4.1. And, for a 600 m grid, if $|\Delta d| = 1500$ m, the mean error is no more than 450 m, which is less than the square size; hence, the predicted position remains constant in accordance with grid-based kNN. As a result, the effect of altitude variation within 1500 m on grid design can be ignored. Contrarily, we present a method to split the grid plane into numerous layers, with the altitude difference within each layer being less than 1500 m.

4.5.2 Similarity Measurement over Ciphertexts

A new method for secure similarity assessment is proposed using VHE, which will play an important role in the encrypted kNN algorithm. To begin, suppose there are three vectors x_1, x_2, and x_3 that are encrypted by VHE into c_1, c_2, and c_3, respectively. The issue is determining how to compare the similarity of three encrypted vectors. Specifically, even without decryption, it is essential to determine which vector between x_1 and x_2 is closer to x_3 in terms of Euclidean distance. To address this issue, the following comparison matrix H is developed. First, the matrix A may be produced by solving the equation $AM = I^*$, where M is a key-switching matrix corresponding to the secret S and I^* is a binary representation of an identity matrix I. Then, H is computed as

$$H = A^T A \tag{4.11}$$

It is important to note that just by H, A cannot be retrieved from H and further get A. The full security analysis will be provided later. As a consequence, even if c and H are known, x is not uniquely determined. Further, we have the following two propositions.

Proposition 4.1 *We have the comparison matrix H, a plaintext x, and the associated ciphertext c, then*

$$c^T H c = w^2 \|x\|^2 \tag{4.12}$$

The correctness of Proposition 4.1 is briefly given below. Moreover, we have $c = M(wx)^*$ and $wx = I^*(wx)^*$, then, we have

$$
\begin{aligned}
c^T H c \\
&= (M(wx)*)^T (A^T A)(M(wx)*) \\
&= ((wx)^*)^T (AM)^T (AM)(wx)^* \\
&= ((wx)^*)^T (I^*)^T (I^*)(wx)^* \\
&= (I^*(wx)^*)^T (I^*(wx)^*) \\
&= (wx)^T (wx) \\
&= w^2 \|x\|^2
\end{aligned}
$$

The above statement suggests a relationship between c, H and $\|x\|^2 (= x^T x)$. It implies that the length of $\|x\|$ may be obtained by simply calculating over ciphertexts. It is worth noting that we only know the length of x but the content of x is unknown.

Proposition 4.2 *Given three plaintexts x_1, x_2, and x_3, the corresponding ciphertexts c_1, c_2, and c_3, and the comparison matrix H, the following condition holds: if $(c_1 - c_3)^T H (c_1 - c_3) \le (c_2 - c_3)^T H (c_2 - c_3)$, then*

$$\|x_1 - x_3\| \le \|x_2 - x_3\|$$

The correctness of Proposition 4.2 may be achieved immediately by using Proposition 4.1 and $VHE.Add$. It is worth noting that Proposition 4.2 has a significant consequence in that we may assess the similarity of vectors without decryption. It is especially clear that such similarity assessment on ciphertexts does not need interactions, reducing transmission costs.

4.5.3 Privacy-Preserving Aircraft Location Verification

PPALV contains the following three phases: *Initialization*, *Offline Training*, and *Online Verification*.

4.5.3.1 Initialization

ATCO configures the system and publishes the system parameters in this phase that includes the steps below.

- *Step1:* In terms of air traffic surveillance needs, ATC first provides the grid parameter $GridParam$, which contains the longitude, latitude, altitude, and size of grid plane, as well as the size of each grid square.
- *Step2:* By calling $VHE.KG(\lambda)$, ATC produces the secret keys S_1 and S_2, as well as the VHE parameters $VHEParam = (l, m, n, p, q, w, \chi)$.
- *Step3:* ATC denotes the kNN algorithm parameter $kNNParam$, which includes the k value and the d_{th} threshold.
- *Step4:* ATC makes public the system parameters ($GridParam$, $VHEParam$, $kNNParam$) and sends S_1 and S_2 to CPS for private storage.

4.5.3.2 Offline Training

CPS conducts offline training on the whole surveillance grid plane in this phase.

- *Step1:* CPS computes the fingerprint for each square in the grid plane $\mathscr{R}$ using the TDOA data received from ADS-B sensors. Assume that $\mathscr{R}$ includes m grid squares and that there are n sensors with detection ranges inside $\mathscr{R}$. Then, for $\mathscr{R}$, there is a data set $D = \{(x_i, y_i), i = 1, \cdots, m\}$, where $x_i \in \mathbb{Z}^n$ is the fingerprint vector of the i-th square with respect to n sensors, and $y_i \in \mathbb{Z}^2$ is the location of two-dimensional coordinates of the i-th square in the plane.
- *Step2:* CPS encrypts D into D_e using the secrets S_1 and S_2 as $D_e = \{(c_i, l_i), i = 1, \cdots, m\}$, where $c_i = VHE.Enc(x_i, S_1)$ and $l_i = VHE.Enc(y_i, S_2)$.
- *Step3:* CPS calculates the comparison matrix H_1 based on S_1 as described in Sect. 4.5.2.
- *Step4:* CPS sends D_e and H_1 to the cloud. It should be noted that the $\{(c_i, l_i), i = 1, \cdots, m\}$ should be sent in a disorderly manner; else, significant security issues would arise. The following is a short explanation: If the published grid parameters are awarded by the cloud server, by using the reverse transmission procedure, the index i can be converted to grid square coordinates.

4.5.3.3 Online Verification

The online verification of PPALV is detailed in the following steps, which is also shown in Fig. 4.9.

- *Step1:* Assume an aircraft is flying into the monitoring area. CPS first calculates the fingerprint x by gathering TDOAs from sensors and then encrypts x with S_1 by executing $c \leftarrow VHE.Enc(x, S_1)$. The encrypted fingerprint c is then transmitted to the cloud server.

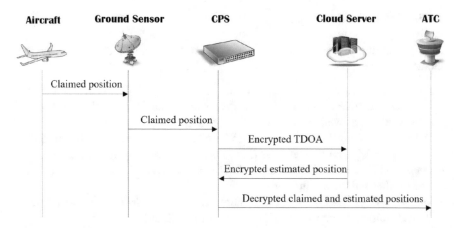

Fig. 4.9 Online verification process

Algorithm 1 Aircraft location estimation
___
Input: D_e, H_1 and c
Output: l'
1: **for** $i = 1$ to m **do**
2:     Calculate $t_i \leftarrow (c - c_i)^T H_1 (c - c_i)$
3: **end for**
4: Find k nearest grids $\{t'_i, i = 1, \cdots, k\}$ by running kNN in $\{t_i, i = 1, \cdots, m\}$
5: Find the corresponding $\{l'_1, l'_2, \cdots, l'_k\}$ in D_e
6: Calculate $l' \leftarrow \frac{1}{k}(l'_1 + l'_2 + \cdots + l'_k)$
7: **return** Encrypted estimated location l'

- *Step2:* CS evaluates the estimated aircraft position over ciphertexts. In specific, by executing the grid-based kNN algorithm, CS gets the k nearest squares $l'_1, \cdots, l'_k$ and averages them as $l' = \frac{1}{k}(l'_1 + l'_2 + \cdots + l'_k)$, in which l' is the encrypted estimated position. CS then sends back l' to CPS. The *Step2* procedure may be reviewed in Alg. 1, and its correctness is guaranteed as

$$(c - c_i)^T H_1 (c - c_i) = w^2 \|x - x_i\|^2$$

- *Step3:* Before gleaning the claimed location y from the received *message*, CPS calls $y' \leftarrow VHE.Dec(l', S_2)$ to obtain the estimated location y'. Finally, y and y' are sent to ATC for aircraft location verification.
- *Step4:* ATC determines if the following equation is true:

$$\|y - y'\| < d_{th} \qquad (4.13)$$

If so, the claimed location is validated and shown on the ATC monitoring system; if not, the received *message* will be permanently deleted.

4.6 Extended Discussion

In this section, we provide a quick method for determining aircraft legality. As mentioned earlier, the grid-based kNN algorithm is used to determine the validity of an aircraft's reporting positions. However, as the number of grid squares increases, so does the computational cost. To enhance the accuracy of air location estimation, for example, for a grid plane of $2°$ longitude by $2°$ latitude, the square size should be set small enough, e.g., $75 \times 75 \, \text{m}^2$. In this instance, the plane has about six million grid squares, and the aircraft's fingerprint must be matched to the fingerprints of such large grids. In the next section, we present a quick aircraft position verification method that uses just a limited number of grids in the computation.

Remember the scheme from Sect. 4.5.3. The estimated position of the aircraft is initially computed using grid-based kNN by calculating encrypted Euclidean distances between the aircraft and all grid squares in the plane. Following that, compare it to the stated location. Finally, if their discrepancy is greater than some specified threshold, d_{th}, the position claim proves fallacious.

It is obvious that the scheme's most time-consuming function is evaluating the estimated position by traversing the whole plane for all grids. Thus, by verifying the location claim but not calculating the aircraft's true position, the amount of computation is minimized. On the basis of these considerations, the following method for fast verification is given. The fundamental thought is to construct a circle with the stated location as the center and a radius of $R_1 = d_{th}$. If the location claim is genuine, it is enclosed in the interior of this circle. Contrarily, the estimated position is in close proximity to the borderline of the circle. As a result, its estimated location is limited to the circumference of this circle, as shown in Fig. 4.10. It is not to say that unlawful aircraft are supposed to be beyond the circle, but that training algorithms are restricted to operating within the circle.

As is well known, the grid-based kNN algorithm estimates the location by averaging the top k grid squares, where k squares are contiguous. As a result, even if the estimated location is contained inside this circle, one or more of the k grids may be located outside it. As a result, in order to maintain accuracy, the training circle

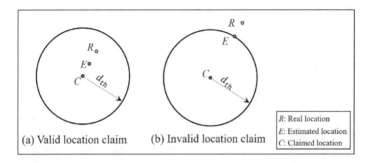

(a) Valid location claim   (b) Invalid location claim

R: Real location
E: Estimated location
C: Claimed location

Fig. 4.10 Small training circle

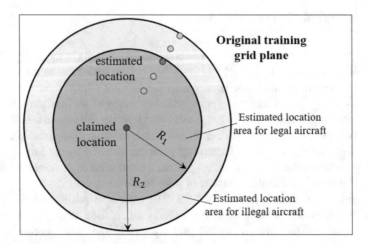

Fig. 4.11 Finding the region of training circle

should be larger to include such squares. For $k = 5$ of the common setting, the greatest distance occurs when five neighboring squares form a straight line, with the midway being the predicted location, which is also at the circle's border, as shown in Fig. 4.11. A larger circle with a radius of $R_2 = d_t h + 2s$ is formed, where s is the grid square side length. As a result, the location claim is valid if the estimated position R falls within the smaller circle ($R < R_1$); if R is contained inside the larger circle ($R_1 < R < R_2$), the location claim is invalid.

In a typical context, the area of a grid plane is $3.3 \times 10^{10} \, \text{m}^2$. However, if $d_{th} = 500 \, \text{m}$ and $s = 75 \, \text{m}^2$, the size of the training circle is only $1.33 \times 10^6 \, \text{m}^2$, which is much smaller than the area in the original plane. As a result, it is much easier to determine the aircraft's validity in the training circle than it is on the whole grid plane. Since a consequence, if actual aircraft locations are not being evaluated, grids beyond the training circle are superfluous for differentiation purposes, as the distances between these grids and the claimed position exceed the threshold. These grids are used in calculations only when ATC wants to determine the true location of an unlawful aircraft.

4.7 Security Analysis

Communication between aviation intranets is secure. In addition, the high-speed dedicated connection also protects the communication between the internal CPS and the external CS, ensuring that the authenticity and integrity of the location data will not be destroyed by attackers. However, the cloud is often considered trustworthy but inquisitive, with incentives to spy on position privacy for economic or other reasons. As a result, security should be ensured for outsourced kNN

computing. We specifically examine the security of PPALV from two perspectives:
(1) secure distance metrics in kNN; (2) confidentiality of aircraft location claims
and fingerprints, as well as grid coordinates and fingerprints.

4.7.1 Confidentiality

We achieve PPALV confidentiality based on VHE. Aircraft position data (location
claims and fingerprints of aircraft) and grid data (coordinates and fingerprints of
grids) are encrypted under the secret keys S_1 or S_2, respectively. The difficult issue
of learning with errors (LWE) ensures the security of VHE. Such confidentiality is
ensured if CPS keeps S_1 or S_2 privately. As a result, the cloud cannot extract valuable information from ciphertexts, thus safeguarding data confidentiality. Besides,
during the offline training phase, encrypted grid coordinates and fingerprints,
$\{(c_i, l_i), i = 1, \cdots, m\}$, have been disordered for transmission. Although the cloud
gains the published grid features, the top K coordinates of the grid cannot be worked
out via the inverse transmission method.

4.7.2 Privacy

Note that x is the fingerprint of the location claim, and encrypted as c; $x_i, i \in$
$\{1, \cdots, m\}$ is the fingerprint of the grid square, and encrypted as c_i. And a
comparison matrix H is calculated as

$$\begin{cases} AM = I^* \\ H = A^T A \end{cases} \tag{4.14}$$

Then c, c_i, and H are uploaded, by CPS, onto the cloud for the distance measurement between c and c_i. The distance

$$\|x - x_i\|^2 \leftarrow \frac{1}{w^2}(c - c_i)^T H(c - c_i)$$

is then calculated by the cloud. Considering location privacy, the above calculations
should not leak any information about x and x_i, which means that any location
fingerprints cannot be directly extracted from c and c_i. However, provided with
H, the cloud has stronger capabilities, which increases the possibility of location-
privacy leakage. Below, we prove the security of fingerprints even if the cloud is
equipped with H access in Theorem 4.1. Besides, we also show the secret-key
security in Theorem 4.2.

Theorem 4.1 *It is infeasible to extract x and x_i from c, c_i and H.*

Proof To simplify, we only consider the infeasibility of restoring x, and the situation is similar for x_i. First, H does not leak information on M; otherwise, if obtaining M from H, the cloud is capable of extracting x from c considering $c = M(wx)^*$. Specifically, in terms of Eq. (4.14), $AM = I^*$ and $H = A^T A$. It seems that it is very likely to restore M, that is, first extract A from $H = A^T A$ and then solve the equation $I^* = AM$. Fortunately, the equation $H = A^T A$ has an infinite number of solutions for A, thus there is infinitely many M. Then we provide a brief analysis as follows.

Obviously, from $I = Q^T Q$, there exists an unbounded number of orthogonal bases $Q \in \mathbb{R}^n$, then

$$A^T A = A^T I A$$
$$= (QA)^T (QA)$$
$$= A^T Q^T QA$$

Theorem 4.2 *Under the assumption that LWE is hard, it is not feasible to extract S even if H is given.*

Proof Assume that given H we can infer M, then due to the equation of $SM = I^* + E$, we can convert the problem of recovering S into solving the hardness problem of LWE. Following this idea, assuming that a solver $\mathscr{A}$ exists for $SM = I^* + E$, then we can derive each element in S and M from $\mathbb{Z}_q$ with $q >> max\{|S|, |M||E|\}$. We further consider $S \in \mathbb{Z}^{m \times n}$ as m row vectors with n dimensions as demonstrated in Eq. (4.15), where $s_j^T (j = 1, 2, \cdots, m)$ is an n-dimensional row vector.

$$S = \begin{bmatrix} s_1^T \\ s_2^T \\ \vdots \\ s_m^T \end{bmatrix}_{m \times n} \tag{4.15}$$

Similarly, $M \in \mathbb{Z}^{n \times ml}$ is treated as ml column vectors with n dimensions as illustrated in Eq. (4.16), where l is a binary representation parameter [27] and $m_j (j = 1, 2, \cdots, ml)$ is an n-dimensional column vector.

$$M = \begin{bmatrix} m_1 & m_2 & \cdots & m_{ml} \end{bmatrix}_{n \times ml} \tag{4.16}$$

Further, $I^* \in \mathbb{Z}^{m \times ml}$ and $E \in \mathbb{Z}^{m \times ml}$ can be rewritten as

$$I^* = \begin{bmatrix} I_{1,1} & I_{1,2} & \cdots & I_{1,ml} \\ \vdots & \vdots & \ddots & \vdots \\ I_{m,1} & I_{m,2} & \cdots & I_{m,ml} \end{bmatrix}_{m \times ml} \tag{4.17}$$

$$E = \begin{bmatrix} E_{1,1} & E_{1,2} & \cdots & E_{1,ml} \\ \vdots & \vdots & \ddots & \vdots \\ E_{m,1} & E_{m,2} & \cdots & E_{m,ml} \end{bmatrix}_{m \times ml} \tag{4.18}$$

From $SM = I^* + E$, we have

$$\begin{cases} s_1^T m_1 = I_{1,1} + E_{1,1} \\ s_1^T m_2 = I_{1,2} + E_{1,2} \\ \vdots \\ s_i^T m_j = I_{i,j} + E_{i,j} \\ \vdots \\ s_m^T m_{ml} = I_{m,ml} + E_{m,ml} \end{cases} \tag{4.19}$$

Therefore, there are $m \times ml$ samples of $(m_j, I_{i,j})$ in

$$I_{i,j} = (s_i)^T m_j - E_{i,j} \tag{4.20}$$

Then solving Eq. (4.20) for s_i^T through $\mathscr{A}$ is equivalent to solving LWE: $b_i = v^T a_i + \varepsilon_i$ for v^T. As a consequence, the hardness of solving LWE is reduced to that of $S'M = I^* + E$. At the beginning of this proof, we have assumed that M can be obtained from H. According to Theorem 4.1, we can get an infinite number of M from H, which also means that an infinite number of S from $SM = I^* + E$.

4.8 Experiment Evaluation

We assess PPALV in this section based on its accuracy, efficiency, and communication overhead.

4.8.1   Experiment Environment

We use a smartphone with a Kirin 1600 MHz microprocessor and 2GB of RAM running Android 4.2.2 to conduct simulations. We also use an Alibaba Cloud graphic workstation equipped with a NVIDIA V100 GPU and 32GB of RAM running CUDA 10.1. Given that a CPS is commonly a resource-constrained device equipped with an ARM processor, the smart phone serves as the CPS in this case. Additionally, the graphic workstation serves as the CS.

   The following considerations are used to establish the system parameters in our experiments. To begin, the surveillance area for $GridParam$ is a grid plane with a longitude of $2°$, a latitude of $2°$, and an altitude of 11000 m that encompasses the aircraft's en route flying phase at cruising altitude. Additionally, the grid square sizes range from 50 m to 600 m to evaluate the efficiency and accuracy of PPALV. We demonstrate the air traffic monitoring region in this section by using Google Earth to display the test environment, as shown in Fig. 4.12. Second, the security level for $VHEParam$ is set to $\lambda = 128$ to provide realistic security. Then, we define $w = 230, l = 50$, and $eBound = 200$ to verify that the ciphertext domain operation is correct. Third, it is worth noting that Strohmeier et al. conducted experiments to demonstrate that the optimum neighbor selection is $k = 5$. As a result, we provide the same $kNNParam$. Besides, we use real-world flight data acquired from

Fig. 4.12 Visualization of test environment

OpenSky, which includes a data collection of over 300,000 ADS-B messages from a two-week sample, with each message being received by at least three sensors [30]. Note that our source code is available at [42].

4.8.2 Accuracy

The technology we used to predict the position of the aircraft can be quantified, and the accuracy of a few hundred meters suffices to verify the aircraft for surveillance. As a result, we begin this part by evaluating the location-estimating approach's efficacy using gathered fight data. Then, we inject false locations to show that our location-validating technique may verify simulated attackers.

4.8.2.1 Accuracy in Estimating Legal Aircraft

The claimed position of a legal aircraft is directly accessible from GNSS and is, therefore, considered valid, and the accuracy of the location estimate is achieved in this section using real-world flight data. We specifically conduct evaluations for plaintexts and ciphertexts on the same datasets. We may also verify the validity of PPALV by comparing these two implementations. The findings of the evaluation are shown in Fig. 4.13. Obviously, reducing the grid square size has a beneficial impact. To be specific, the smaller square size will bring on more accurate estimation. For instance, reducing the grid square size from 600 m to 50 m also results in a reduction in mean errors of up to 29.22%. Since actual data contains noise, the 99th percentile measure is more sensitive to outliers than the mean error, median error, or root mean squared error (RMSE). For example, the horizontal discrepancy between the actual and predicted positions on a 600 m grid is 390 m. As a result, we set $d_{th} = 500$ m threshold in $kNNParam$ for future location validation. More significantly, PPALV achieves about the same level of accuracy as the standard plaintext method. However, while our solution provides significant privacy advantages that are necessary to securing aircraft's position verification in the cloud environment, the plaintext approach does not currently guarantee this kind of protection.

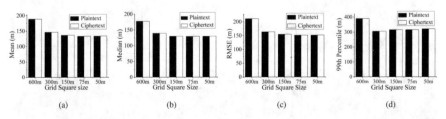

Fig. 4.13 Accuracy of estimating legal aircraft. (**a**) Mean error. (**b**) Median error. (**c**) RMSE. (**d**) 99th percentile

By extending the Kalman filter, Zhao et al. recently gave an ADS-B positioning scheme based on TDOA/AOA, which improves the positioning accuracy and monitoring robustness [43]. For a grid square of 600 m, this scheme achieves the RMSE error 650 m but AEALV only 210 m. In addition, this scheme does not encrypt the outsourced location data, so the location privacy of the aircraft cannot be achieved. In contrast, AEALV not only guarantees better estimation accuracy but also preserves such privacy.

4.8.2.2 Accuracy in Detecting Illegal Aircraft

We first mimic a single attacker transmitting fictitious locations. Specifically, we begin by extracting an actual track made up of 500 locations from the OpenSky data and then comparing it to a simulated assault scenario. We analyze the OpenSky data under the ADS-B message protocol, change the field holding positional information, and recreate the parity-check value [4]. Notably, this simulated data conforms to the ADS-B message structure and can be read by ADS-B transponders, giving the appearance of coming from a genuine aircraft. These bogus signals follow a particular geographical pattern in which the actual and fake tracks begin in the same place but then deviate at an arbitrary angle of 5 to 30°. In Google Earth, Fig. 4.14 displays such two trajectories, with the red line representing a genuine track and the green line representing a fabricated one.

Following that, we examine whether or not the false locations can be confirmed using the fast verifying method mentioned in Sect. 4.6. We compute the attacker's TDOA values first and then evaluate the detection rate by testing the attack scenario 500 times. Table 4.2 summarizes the results of our location verification method. As can be seen, it achieves the same *Accuracy*, *Recall*, and *Precision* as the

Fig. 4.14 Real and fake trajectories

Table 4.2 Accuracy of detecting illegal aircraft

Threshold [m]	500		600		700		800	
	Plaintext	Ciphertext	Plaintext	Ciphertext	Plaintext	Ciphertext	Plaintext	Ciphertext
Accuracy	1	1	1	1	0.93	0.93	0.5	0.5
Recall	1	1	1	1	1	1	1	1
Precision	1	1	1	1	0.87	0.87	0.5	0.5

plaintext method for detecting unauthorized aircraft. For the classification performance indicators, $Precision = \frac{TP}{TP+FP}$, $Recall = \frac{TP}{TP+FN}$ and $Accuracy = \frac{TP+TN}{TP+TN+FP+FN}$, where TP is *True Positive*, TN is *True Negative*, FP is *False Positive* and FN is *False Negative* [44]. These indications naturally diminish when thresholds are increased. The *Accuracy* and *Precision* values, fall by 50% at the 800 m barrier. As a result, the 500 m threshold is an excellent option for ensuring the detection rate.

4.8.3 Efficiency

The efficiency of location verification is essential to ensure flight safety in air traffic surveillance. Below we evaluate the efficiency of PPALV in detail.

The stages of *Initialization* and *Offline Training* are completed prior to *Online Verification*, and, therefore, they have no effect on the efficiency of ALV. The efficiency of *Online Verification* is time-dependent, mainly including the time required by CPS to encrypt the aircraft location claim and fingerprints, the time required by the cloud to implement the grid-based kNN algorithm, and the round-trip time of the transmission between the cloud and CPS. Since the transmission is carried out through a dedicated line, the time required is negligible, so the following only analyzes the CPS encryption time and the cloud execution time.

CPS Encrypting Time As mentioned earlier, when an aircraft goes into the monitoring area, CPS will encrypt the fingerprint $x \in \mathbb{Z}^n$ through calling $c \leftarrow VHE.Enc(x, S_1)$. It is, therefore, necessary to measure the VHE encryption time of the n-dimensional vector. In the monitoring grid plane, the number of sensors rarely exceeds 10, while if an ADS-B message is received at most 2 sensors, the TDOA-based method is not applicable; thus, $3 \leq n \leq 10$ can be specified. Meanwhile, since CPS is commonly an embedded device with limited resources (e.g., AX680 produced by THALES [45]), the encryption time is evaluated by simulating a smartphone with a Kirin-910 microprocessor running the Android 4.2 operating system.

Specifically, we rebuilt the VHE encryption program using Java 1.7.0 and NDK R9, and then installed it on the smartphone. In addition, by obtaining the root permission of the Android system and running an APP called SetCPU, the processor frequency representing the computing power can be adjusted from 1596 MHz to 208 MHz. Figure 4.15 shows the encryption time versus processor frequency and vector dimension. As shown, the encryption time increases as the frequency drops or the dimension rises. Even at the frequency of 208 MHz, the time required to encrypt a ten-dimensional vector is only 136.09 ms. As a result, VHE encryption has high efficiency, suitable for resource-constrained CPS.

As stated before, when an aircraft enters the specified surveillance area, CPS encrypts the fingerprint $x \in \mathbb{Z}^n$ by invoking $c \leftarrow VHE.Enc(x, S_1)$. As a result, the VHE encryption time of a n-dimensional vector must be determined. Typically,

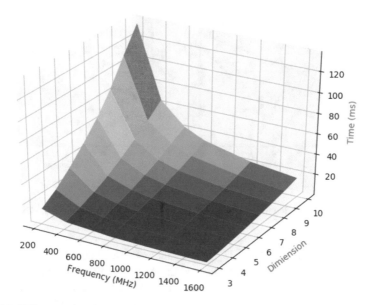

Fig. 4.15 CPS encrypting time

the surveillance grid plane has no more than 10 sensors. However, if only two sensors receive the same ADS-B data, the TDOA technique is inapplicable. As a result, we set $3 \leq n \leq 10$. Simultaneously, given that CPS is often a resource-constrained device, such as the AX680 series supplied by THALES [45], we assess such encryption time using a smart phone equipped with a Kirin 910 ARM processor running the Android 4.2.2 operating system. In particular, we use Java 1.7.0 and the NDK R9 development kit to rebuild the VHE encryption application and subsequently install it on the smart phone. By gaining root access to the Android system, we may change the processor frequency to match calculation capability by launching APPs, such as SetCPU. As shown in Fig. 4.15, the frequencies are varied between 1596 and 208 MHz, and the encryption becomes more inefficient as the frequency or dimension decreases. Even at the lowest frequency of 208 MHz, the time required to encrypt a ten-dimensional vector is just 136.09 ms. As a result, VHE encryption is much more efficient and, therefore, better suited for resource-constrained CPS.

Cloud Running Time On the cloud side, the primary time sink is computing $((c - c_i)^T H_1 (c - c_i)$ for large ciphertexts, for which the number of ciphertexts may reach six million when using a grid size of 75 m. To ensure the efficiency of these computations, we use an Alibaba Cloud server equipped with a NVIDIA TESLA V100 GPU and 32 GB RAM, and running CUDA 10.1. The V100 contains 5120 CUDA cores and can handle large-scale parallel workloads. The computation of $(c - c_i)^T H_1 (c - c_i)$ with millions of elements is inherently parallelable. As a result, it is an excellent candidate for GPU acceleration. The next section evaluates the cloud's performance using six million data points.

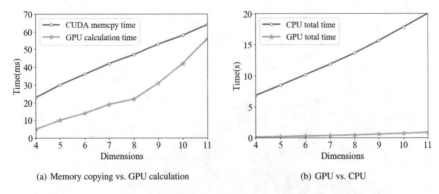

(a) Memory copying vs. GPU calculation   (b) GPU vs. CPU

Fig. 4.16 GPU computational time. (**a**) Memory copying vs. GPU calculation. (**b**) GPU vs. CPU

The calculation of $(c - c_i)^T H_1 (c - c_i)$ includes two operations, vector-matrix multiplication and vector subtraction, and the calculation time of the latter can be ignored relative to the former. Hence such the calculation can be reformulated as $v_i^T M v_i$, in which $v_i, i = 1, \cdots, n$ is an m-dimensional vector and M is an $m \times m$ matrix. In order to perform this calculation in parallel using CUDA, we first allocate contiguous memory space for m-dimensional n data items and replicate these data from the CPU memory to the GPU global memory. Following that, $v_i^T M v_i$ is run in the single instruction multiple threads (SIMT) mode. Specifically, we treat $M \in \mathbb{R}^{m \times m}$ as $M = [M_1, \cdots, M_m]$, in which M_j is the jth column of M. By doing so, $v_i^T M$ is split into m vector-vector multiplications as $v_i^T M_j, j = 1, \cdots, m$, thus implementing high parallelism. Finally, we copy the results from the GPU global memory to the CPU memory.

Figure 4.16a depicts the time of CUDA memory copying compared to that of GPU calculation. It is very intuitive to see how they grow. The former has a trend of linear growth, while the latter grows quadratically. For low-dimensional vectors, the former is often longer than the latter. The delay becomes more noticeable when memory copying happens many times during a single computation. As a result, we use the page-locked method to speed up memory copying. Keep in mind that when the dimension of 8 grows, the time required for GPU computation increases somewhat. This is because of the restriction of the data organization structure in the GPU; if the dimension is equal to or over 8, the number of warps in a block is likely to eclipse the maximum number of streaming multiprocessor operations, leading to a faster rise in computing time.

Finally, we compare the overall time used by the CPU and GPU (including copying and computation). As shown in Fig. 4.16b, computing on the GPU is much quicker than computing on the CPU for the identical job. Even with 11-dimensional data items, the total time required by the GPU is 683 ms (617 ms for CUDA memory transfer and 56 ms for GPU computation), which is much less than the time required by the CPU (over 20 s), just around $\frac{1}{30}$. As a result, PPALV is a viable method for aircraft position verification.

4.8.4 Communication Cost

We evaluate the communication costs solely during the *Online Verification* phase, since they may influence validating efficiency. These types of communication comprise two components. To begin, CPS uploads the aircraft's encrypted fingerprint to the cloud; after that, the cloud provides the aircraft's encrypted estimated position to CPS. because the entire quantity of data sent is less than 1 KB, the communication cost associated with the high-speed dedicated line is negligible.

4.9 Conclusion

We have proposed PPALV for secure aircraft location verification in ADS-B in this chapter, using the grid-based kNN, which makes use of the powerful cloud server to handle large training grid squares and thus improves the accuracy of aircraft location verification. Security analysis shows that PPALV protects the privacy of aircraft location and grid data. Besides, we have designed a rapid method for determining the authenticity of aircraft locations, substantially reducing the time required for location verification. Performance assessment shows that PPALV has high efficiency not only on embedded ARM processors but also on cloud servers. Even with six million grid cells, the time required to evaluate encrypted distances is less than 200 ms, including GPU calculation time and CUDA memory copying time. We will continue to investigate more ADS-B security issues in the future, e.g., physical layer intrusion detection.

References

1. C. Ashton, A. S. Bruce, G. Colledge, and M. Dickinson, "The search for mh370," *The Journal of Navigation*, vol. 68, no. 1, pp. 1–22, 2015.
2. T. L. Davis, "Global sesar1/NextGen internet based ATN infrastructure," in *Proceedings of IEEE ICNS 2016*, 2016, pp. 1–17.
3. M. Strohmeier, M. Schäfer, R. Pinheiro, V. Lenders, and I. Martinovic, "On perception and reality in wireless air traffic communication security," *IEEE transactions on intelligent transportation systems*, vol. 18, no. 6, pp. 1338–1357, 2016.
4. H. Yang, Q. Zhou, M. Yao, R. Lu, H. Li, and X. Zhang, "A practical and compatible cryptographic solution to ads-b security," *IEEE Internet of Things Journal*, vol. 6, no. 2, pp. 3322–3334, 2019.
5. M. Schäfer, V. Lenders, and J. Schmitt, "Secure track verification," in *Proceedings of IEEE S&P 2015*, 2015, pp. 199–213.
6. D. Moser, P. Leu, V. Lenders, A. Ranganathan, F. Ricciato, and S. Capkun, "Investigation of multi-device location spoofing attacks on air traffic control and possible countermeasures," in *Proceedings of ACM MobiCom 2016*, 2016, pp. 375–386.
7. K. Jansen, M. Schäfer, D. Moser, V. Lenders, C. Pöpper, and J. Schmitt, "Crowd-GPS-Sec: Leveraging crowdsourcing to detect and localize GPS spoofing attacks," in *Proceedings of IEEE S&P 2018*, 2018, pp. 1018–1031.

8. S. Sciancalepore and R. Di Pietro, "SOS: Standard-compliant and packet loss tolerant security framework for ADS-B communications," *IEEE Transactions on Dependable and Secure Computing*, 2019, https://doi.org/10.1109/TDSC.2019.2934446.

9. M. Strohmeier, V. Lenders, and I. Martinovic, "On the security of the automatic dependent surveillance-broadcast protocol," *IEEE Communications Surveys and Tutorials*, vol. 17, no. 2, pp. 1066–1087, 2015.

10. K. Zheng, H. Wang, H. Li, W. Xiang, L. Lei, J. Qiao, and X. S. Shen, "Energy-efficient localization and tracking of mobile devices in wireless sensor networks," *IEEE Transactions on Vehicular Technology*, vol. 66, no. 3, pp. 2714–2726, 2016.

11. S. B. Cruz, T. E. Abrudan, Z. Xiao, N. Trigoni, and J. Barros, "Neighbor-aided localization in vehicular networks," *IEEE Transactions on Intelligent Transportation Systems*, vol. 18, no. 10, pp. 2693–2702, 2017.

12. J. A. Belloch, J. M. Badía, F. D. Igual, and M. Cobos, "Practical considerations for acoustic source localization in the IoT era: Platforms, energy efficiency, and performance," *IEEE Internet of Things Journal*, vol. 6, no. 3, pp. 5068–5079, 2019.

13. D. Vasisht, S. Kumar, and D. Katabi, "Decimeter-level localization with a single WiFi access point," in *Proceedings of USENIX NSDI 2016*, 2016, pp. 165–178.

14. Y. A. Nijsure, G. Kaddoum, G. Gagnon, F. Gagnon, C. Yuen, and R. Mahapatra, "Adaptive air-to-ground secure communication system based on ADS-B and wide-area multilateration," *IEEE Transactions on Vehicular Technology*, vol. 65, no. 5, pp. 3150–3165, 2015.

15. M. Strohmeier, I. Martinovic, and V. Lenders, "A k-NN-based localization approach for crowdsourced air traffic communication networks," *IEEE Transactions on Aerospace and Electronic Systems*, vol. 54, no. 3, pp. 1519–1529, 2018.

16. M. Mazzocchi, F. Hansstein, and M. Ragona, "The 2010 volcanic ash cloud and its financial impact on the European airline industry," in *Proceedings of CESifo 2010*, vol. 11, no. 2, 2010, pp. 92–100.

17. "Aviation cloud," http://www.aviationcloud.aero/, accessed November 20, 2019.

18. B. Butler, "How Boeing is using the cloud," https://www.networkworld.com/article/2175805/how-boeing-is-using-the-cloud.html/, accessed April 3, 2014.

19. K. Sampigethaya, R. Poovendran, and C. S. Taylor, "Privacy of general aviation aircraft in the NextGen," in *Proceedings of IEEE DASC 2012*, 2012, pp. 7B5–1.

20. T. Larsen, "Cross-platform aviation analytics using big-data methods," in *Proceedings of IEEE ICNS 2013*, 2013, pp. 1–9.

21. H. Li, D. Liu, Y. Dai, T. Luan, and S. Yu, "Personalized search over encrypted data with efficient and secure updates in mobile clouds," *IEEE Transactions on Emerging Topics in Computing*, vol. 6, no. 1, pp. 97–109, 2018.

22. Y. Zhang, C. Xu, H. Li, K. Yang, J. Zhou, and X. Lin, "HealthDep: An efficient and secure deduplication scheme for cloud-assisted ehealth systems," *IEEE Transactions Industrial Informatics*, vol. 14, no. 9, pp. 4101–4112, 2018.

23. J. Liang, Z. Qin, S. Xiao, L. Ou, and X. Lin, "Efficient and secure decision tree classification for cloud-assisted online diagnosis services," *IEEE Transactions on Dependable and Secure Computing*, 2019, doi: 10.1109/TDSC.2019.2922958.

24. A. Yang, J. Xu, J. Weng, J. Zhou, and D. S. Wong, "Lightweight and privacy-preserving delegatable proofs of storage with data dynamics in cloud storage," *IEEE Transactions on Cloud Computing*, 2018, doi: 10.1109/TCC.2018.2851256.

25. P. Paillier and D. Pointcheval, "Efficient public-key cryptosystems provably secure against active adversaries," in *Proceedings of Springer ASIACRYPT 1999*, 1999, pp. 165–179.

26. C. Gentry, "Fully homomorphic encryption using ideal lattices," in *Proceedings of ACM STOC 2009*, 2009, pp. 169–169.

27. H. Zhou and G. Wornell, "Efficient homomorphic encryption on integer vectors and its applications," in *Proceedings of IEEE ITA 2014*, 2014, pp. 1–9.

28. M. Strohmeier, V. Lenders, and I. Martinovic, "Lightweight location verification in air traffic surveillance networks," in *Proceedings of ACM CPSS 2015*, 2015, pp. 49–60.

29. D. Demmler, T. Schneider, and M. Zohner, "Aby-a framework for efficient mixed-protocol secure two-party computation," in *Proceedings of NDSS 2015*, 2015, pp. 1–15.
30. Opensky, "Open air traffic data for research," https://opensky-network.org/, accessed October 02, 2019.
31. S. Zhang, H. Li, Y. Dai, J. Li, M. He, and R. Lu, "Verifiable outsourcing computation for matrix multiplication with improved efficiency and applicability," *IEEE Internet of Things Journal*, vol. 5, no. 6, pp. 5076–5088, 2018.
32. M. Gruteser and D. Grunwald, "Anonymous usage of location-based services through spatial and temporal cloaking," in *Proceedings of ACM MSAS 2003*, 2003, pp. 31–42.
33. M. Andrés, N. Bordenabe, K. Chatzikokolakis, and C. Palamidessi, "Geo-indistinguishability: Differential privacy for location-based systems," in *Proceedings of ACM CCS 2013*, 2013, pp. 901–914.
34. G. Xu, H. Li, Y. Dai, K. Yang, and X. Lin, "Enabling efficient and geometric range query with access control over encrypted spatial data," *IEEE Transactions on Information Forensics and Security*, vol. 14, no. 4, pp. 870–885, 2018.
35. D. Liu, A. Alahmadi, J. Ni, X. Lin, and X. Shen, "Anonymous reputation system for IIoT-enabled retail marketing atop pos blockchain," *IEEE Transactions on Industrial Informatics*, vol. 15, no. 6, pp. 3527–3537, 2019.
36. Y. Chen, X. Zhu, and S. Gong, "Person re-identification by deep learning multi-scale representations," in *Proceedings of IEEE ICCV 2017*, 2017, pp. 2590–2600.
37. H. Zhu, R. Lu, C. Huang, L. Chen, and H. Li, "An efficient privacy-preserving location-based services query scheme in outsourced cloud," *IEEE Transactions on Vehicular Technology*, vol. 65, no. 9, pp. 7729–7739, 2015.
38. M. Ghaffari, N. Ghadiri, M. H. Manshaei, and M. S. Lahijani, "A peer-to-peer privacy preserving query service for location-based mobile applications," *IEEE Transactions on Vehicular Technology*, vol. 66, no. 10, pp. 9458–9469, 2017.
39. I. Romani de Oliveira, J. Magalhaes Junior, B. Leão, and A. Conto, "Assessing coverage of a dynamic airborne surveillance network for air traffic," 09 2017.
40. J. Sun, X. Olive, M. Strohmeier, M. Schäfer, I. Martinovic, and V. Lenders, "OpenSky report 2021: Insights on ads-b mandate and fleet deployment in times of crisis," *network*, 2021.
41. M. Strohmeier, I. Martinovic, and V. Lenders, "Ak-NN-based localization approach for crowdsourced air traffic communication networks," *IEEE Transactions on Aerospace and Electronic Systems*, vol. 54, no. 3, pp. 1519–1529, 2018.
42. Q. Zhou, "Secure aircraft location verification," https://github.com/qxzhou1010/SecureAircraftLocationVerification, accessed October 02, 2019.
43. D. Zhao, J. Sun, and G. Gui, "En-route multilateration system based on ADS-B and TDOA/AOA for flight surveillance systems," in *Proceedings of IEEE VTC 2020*, 2020, pp. 1–6.
44. C. Sammut and G. I. Webb, *Encyclopedia of machine learning*. Springer Science & Business Media, 2011.
45. "Ax680: Ads-b (automatic dependant surveillance – broadcast)," https://www.thalesgroup.com/en/ax680-ads-b-automatic-dependant-surveillance-broadcast/, accessed November 20, 2019.

Chapter 5
Complete ADS-B Security Solution

5.1 Introduction

We propose a complete cryptographic solution for ADS-B security in this chapter, ensuring privacy and integrity without modifying existing ADS-B protocols or upgrading current ADS-B transponders, thereby being well-suited to large-scale and low-cost deployments. The main idea of our solution is to separate the aircraft identifier from its associated geographic locations to guarantee the privacy of general aviation, maximize the use of reserved fields in the ADS-B message format to protect the integrity of messages, and reduce package loss and disorder for increasingly congested 1090ES data links. Specifically, our contributions are summarized below.

– We offer a comprehensive cryptographic solution for ADS-B communication, which can protect the ADS-B system from active and passive attacks. Security analysis shows that our solution guarantees the confidentiality, authenticity, and integrity of ADS-B communication. In addition, by adapting cryptographic primitives to the ADS-B message format, our solution achieves high compatibility with the existing ADS-B system.
– We use data from OpenSky, a large-scale sensor network designed to collect large amounts of ADS-B data. Simulations on embedded devices and desktop computers demonstrate the efficiency of our solution.
– We illustrate the implementation of our solution in a real airport, where the total time required for authenticating 4500 messages is only 81 ms. Besides, regardless of whether the encryption function is turned on, the monitoring console may still display the same flight trajectory, which proves the feasibility and compatibility of our solution in the real-world ADS-B system.

The remainder of this chapter is structured as follows. Section 5.2 contains references to related work. In Sect. 5.3, we define the secure ADS-B communication problem in detail, including the system model, threat model, and design objectives.

H. Yang et al., *Secure Automatic Dependent Surveillance-Broadcast Systems*,
Wireless Networks, https://doi.org/10.1007/978-3-031-07021-1_5

In Sect. 5.4, we elaborate on the proposed solution. In Sect. 5.5, we give extended discussions. In Sects. 5.6 and 5.7, we evaluate the security and performance of our solution, respectively. In Sect. 5.8, we analyze the compatibility of our solution by deploying it in a real-world airport. In Sect. 5.9, we conclude this chapter.

5.2   Related Work

In order to secure ADS-B communication, cryptographic approaches can be explored to preserve the secrecy and integrity of air traffic data [1–4]. Sampigethaya et al. first exploited the symmetric-key encryption and digital signature to preserve the privacy and authenticity of ADS-B messages [3, 5].

Along the study route, based on a block cipher, Chen et al. presented an authenticated encryption solution to protect the ADS-B communication security [6]; however, it is difficult to manage and distribute symmetric keys in large-scale, dynamic, and not well-connected ADS-B networks. For the ADS-B scenario, Wesson et al. further analyzed the inherent shortcomings of symmetric-key methods, strongly recommending protecting the integrity of ADS-B data through public-key technology [7]. Researchers then proposed a variety of PKI-based ADS-B authentication techniques in succession [8–11]. For instance, Pan et al. [12] made use of X.509 certificates and ECDSA signatures to ensure the integrity of ADS-B messages; unfortunately, because of high costs for the certificate-chain transmission and verification, this PKI-based signature is not suitable for low-bandwidth ADS-B data links.

For efficient ADS-B authentication, the IBS technology is considered to be an alternative method in which the public key can be derived directly from the user's identity (such as email address, phone number, etc.). In this case, a PKI is not required, so the cost of operating and maintaining the PKI can be significantly reduced [13]. However, the existing IBS schemes, such as [14, 15], still need to attach the IBS signature to the sent message, so the existing ADS-B protocol needs to be modified, which may cause incompatibility issues and hinder the deployment of ADS-B in a real-world aviation scene. As we all know, TESLA is considered a lightweight authentication protocol that is commonly adopted in wireless broadcast communication internationally. Nevertheless, due to the special requirements of ADS-B, the straightforward use of TESLA does not seem to be a good method. For example, when urgent messages require real-time authentication, TESLA does not work well because its core idea is based on delayed authentication. Therefore, we have to delve deep into the application of TESLA in the ADS-B communication environment.

As far as the confidentiality protection of ADS-B messages is concerned, asymmetric encryption requires PKI, which brings high costs for certificate-chain computing and communication. Although researchers have also proposed various advanced symmetric ciphers to preserve the privacy of these messages by encrypting the entire ADS-B message, participants without the keys cannot publicly access the encrypted location data, which may affect flight safety [8, 16, 17]. Besides, traditional block ciphers, such as AES, require additional padding to fit a fixed block size, which greatly expands the message length, thereby increasing the burden on the already congested 1090ES data link.

As a consequence, to improve the ADS-B communication security, it is essential to develop privacy- and integrity-preserving ADS-B security schemes with high performance and compatibility.

5.3 Problem Statement

This section first defines the system model, then presents the threat model by identifying different types of attacks, and finally gives design objectives.

5.3.1 System Model

As the backbone of the next-generation ATC system, ADS-B is mainly composed of two data links: UAT and 1090ES. Our research focuses on the latter, which uses a 112-bit data frame and is available in the airspace of most countries worldwide. Our method can be easily extended to UAT.

As shown in Fig. 5.1, our system model abstracts the basic components of a typical ADS-B system, in which the aircraft uses global navigation satellite systems such as GPS to get its position. Then, the transponder sends a continuous ADS-B signal to the surrounding area through ADS-B Out (an air communication mechanism). If the adjacent aircraft and ground stations are installed with ADS-B receivers, they can receive these signals through the related communication subsystem ADS-B In, improving situational awareness. The air traffic controller (ATCO) connected to the ground station will display the aircraft on the monitoring screen based on the received position claim, performing critical monitoring functions. Because of the interconnectivity of civil aviation intranet networks, this study assumes that ground communications are secure.

Then, as the authority of key management, a trusted third party (TTP) must be integrated into the system model. ATCO or another legal entity can act as a TTP to perform the generation and distribution of system parameters and keys. By doing so, we can also assume that there is a secure channel between the TTP and the aircraft for the transmission of parameters and keys. In the real world, the CPDLC can guarantee the secure channel that supports air-to-ground data transmission [2].

It is worth emphasizing that our model mainly considers the security impact of ADS-B on general aviation aircraft, rather than conventional commercial aircraft operating on public and controlled routes.

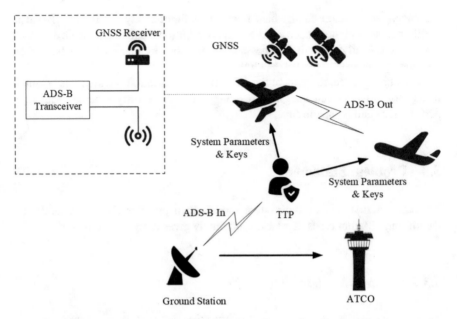

Fig. 5.1 System model of proposed solution

Fig. 5.2 Threat model of proposed solution

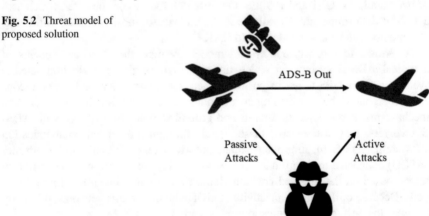

5.3.2 Threat Model

As mentioned earlier, the lack of security primitives in the ADS-B protocol poses major security threats to air traffic because all ADS-B broadcast messages are sent over unencrypted radio broadcasts and no message authentication mechanism is considered. In this part, we present a threat model, as shown in Fig. 5.2, which can capture two different types of actual attacks performed by skilled attackers: passive attacks and active attacks.

- *Passive Attacks*: The FAA strongly recommends the use of clear-text data links to maintain open access to critical surveillance data (such as aircraft location information) to improve flight safety and international interoperability. However, because of the lack of confidentiality, this exposes the ADS-B system to security risks. For example, an attacker would merely need to be equipped with a low-cost ADS-B receiver to intercept broadcast and eavesdrop messages. Here, valuable information, such as the locations of the aircraft, can be associated with their identities, leading to privacy leaks.
- *Active Attacks*: Attackers may forge flight data that fully conforms to the ADS-B message format and then use commercial and low-cost ADS-B transmitters to broadcast it, leading to the so-called Ghost Aircraft Injection attack. Due to the lack of authentication and integrity measures in the ADS-B protocol, an aircraft that receives a forged signal may maneuver to avoid a collision with a non-existent aircraft. Moreover, erroneous information injection can significantly impair air traffic management by placing large numbers of ghost aircraft on ATCO displays, thereby limiting potential operational capabilities. In addition, the bit flip signal of the input physical channel will definitely cause the existing plane to be removed from the monitoring screen [1].

5.3.3   Design Objectives

With reference to the above models, the overall goal of the proposed solution is to provide a comprehensive cryptographic solution for ADS-B security, which is compatible and high performance, and can successfully defend against passive attacks and active attacks. To this end, the solution should satisfy the following security objectives.

- *Privacy*: Unauthorized entities should not establish a link between the aircraft's digital identity and important data (such as extremely precise location information). Especially for private jets owned by companies or individuals, their geographic trajectories are more likely to be associated with business or personal travel. Therefore, the anonymity of identification should also be established to preserve the privacy of general aviation aircraft.
- *Authenticity and integrity*: ADS-B data received must be validated to eliminate malicious infusions that could fool air traffic surveillance and collision avoidance systems.
- *Robustness to packet loss and disorder*: Considering that the required rollout of ADS-B may lead to an increase in channel usage in the next few decades, the loss and disorder of ADS-B data packets on the physical layer may increase significantly. Thus, our solution is supposed to successfully deal with packet loss and disorder.

- *Efficiency*: Our solution should be low-overhead in terms of communication and computing, taking into account low-bandwidth data links and resource-constrained avionics to alleviate expensive system requirements.
- *Compatibility*: To be used in the current system, our solution must provide high compatibility without modifying the existing ADS-B system, which is critical in view of the long approval and acceptance cycle associated with upgrading existing aviation technology.

5.4   Proposed Solution

As illustrated in our threat model, there are two types of potential threats to ADS-B systems: passive attacks that compromise privacy by eavesdropping on aircraft positions, velocities, etc. and associating them with aircraft identities; and active attacks that break integrity and authenticity by injecting forged messages or corrupting traffic data to confuse air traffic surveillance systems. In this part, we outline our fundamental strategy for defending against these assaults. First, Table 5.1 gives the commonly used notations in this chapter.

5.4.1   Resistance to Passive Attacks

We merely encrypt the AA field of the ADS-B message to prevent passive attacks. Three benefits are listed here. First, the unique $ICAO$ address carried in the AA field is used to identify the aircraft. Encrypting the $ICAO$ address can prevent attackers from associating the geographic location of the aircraft with its real digital ID, preventing attackers from identifying specific aircraft and ensuring the privacy

Table 5.1 Notations of proposed solution

Notation	Meaning
λ	Security parameter
$\mid \cdot \mid$	Bit length
$\parallel$	Concatenation operator
T	FFX tweak
K_F	FFX key
$FFX.Encrypt_K^T(\cdot)$	FFX encryption
K_i	The ith key
P_i	The ith packet
Γ_i	The ith authenticating code
$F(\cdot)$	One-way function
$F'(\cdot)$	Truncation function
pid	Encrypted $ICAO$

of aircraft recognition. Second, when selecting appropriate encryption primitives to encrypt $ICAO$, traditional block ciphers such as AES require a fixed block size, such as 64 or 128 bits, so 24-bit $ICAO$ must be filled with extra bits; because of the ADS-B message format, padding is not possible. In this section, we use FFX as an underlying encryption primitive that can support plaintext of any length. In addition, the advantage of preserving the format in FFX is that $ICAO$ and its related ciphertext can keep the same length and alphabet, making it compatible with the current ADS-B protocol. Third, encrypting the entire ADS-B message will violate the public availability of location data, leading other participants (such as neighboring aircraft and ground stations) to be unaware, and may reduce airspace surveillance capabilities. Our technology only encrypts AA field to make other aircraft data clear and accessible without affecting flight safety.

In short, a TTP is incorporated into our solution to enable FFX encryption when receiving $ICAO$ from the registered aircraft. In this case, encryption is entrusted to TTP instead of avionics with limited computing power, thereby reducing the load on ADS-B equipment. We replace the $ICAO$ address of the aircraft with pid, which is the corresponding $ICAO$ ciphertext obtained from TTP. Figure 5.3 illustrates the encryption and replacement process. In the later sections, we will delve into how to incorporate this encryption technology into our solution.

5.4.2 Resistance to Active Attacks

In order to avoid malicious injection by active attackers, we use the one-way key chain and keyed MAC to make TESLA properly adapt to the complex electromagnetic environment of ADS-B communication. In addition, the lack of a data packet re-transmission mechanism in the ADS-B protocol may cause frequent packet losses at the physical layer. As shown in [18], the average packet error rate

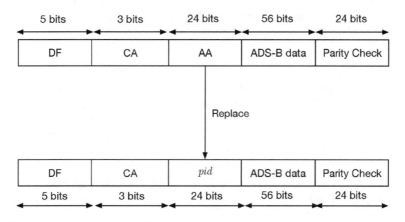

Fig. 5.3 Resistance to passive attacks

is as high as 33%, showing that ADS-B data connections have obvious packet loss. We will use the reserved ADS-B field to store the key and MAC and send them with the message to be verified. Here, even if the ADS-B protocol and avionics remain unchanged, the original message may still be correctly parsed, and all ADS-B functions are kept. Therefore, our method can be combined with the existing ADS-B protocol. Below, we will elaborate on the basic broadcast authentication process in the ADS-B scenario, and the complete implementation method will be introduced in the following part.

Sender To broadcast the ith ADS-B message Ω_i, the sender firstly chooses K_{i-1} and K_i in the *Keychain*, concatenates the ADS-B Data M_i and K_{i-1} together forming D_i, calculates the keyed-hash message authentication code Γ_i as $H(K_i, D_i)$, and appends Γ_i to D_i generating the packet P_i to be sent. Finally, P_i is broadcast through ADS-B Out. Figure 5.5 illustrates the procedure of generating the ADS-B packet.

Receiver When receiving P_i via ADS-B-In, based on the delayed-authentication idea, the receiver does not instantly check the validity of P_i because of the unawareness of K_i that is essential to generate Γ_i. Here, the receiver first buffers P_i until P_{i+1} of containing the K_i reaches to extract K_i to generate the authentication code Γ_i'. The receiver then completes packet verification by comparing Γ_i' with Γ_i retrieved from buffered P_i. The procedure of verifying ADS-B packets is shown in Fig. 5.6.

5.4.3   Framework

By incorporating the above security countermeasures of resisting passive and active attacks, we come up with a complete framework to show how privacy and integrity can be guaranteed, simultaneously not requiring the alteration of the existing ADS-B protocol. Our solution is composed of four PPT algorithms as $ParamGen(\Theta)$, $KeyChainGen(F)$, $Encrypt(E)$, and $Authenticate(H)$.

- $\Theta(\lambda)$. Taking the security level λ as input, TTP issues the system parameters (e.g., T, K_F, n).
- $F(K_n)$. TTP recursively runs the one-way function $F(\cdot)$ to calculate *Keychain* as $F^v(K_n) = F(F^{v-1}(K_n))$, in which *Keychain* $= (K_0, K_1, ..., K_n)$. In our solution, we instantiate F as MD5 meaning $\{|K_i| = 128|0 \leq i \leq n\}$.
- $E(ICAO, K_F, T)$. Taking the aircraft's $ICAO$, the key K_F and the tweak T as input, TTP invokes $FFX.Encrypt_K^T(ICAO)$ that then outputs pid, the ciphertext of $ICAO$, to the aircraft.
- $H(K_i, D_i)$. The aircraft generates Γ_i for D_i through the keyed-hash function H with the key K_i. In our proposal, we implement H using HMAC-MD5-96 implying $\{|\Gamma_i| = 96|1 \leq i \leq n\}$.

Note that we utilize the truncation function $F'(K_i)$ to shorten the key from $\{|K_i| = 128 | 0 \le i \le n\}$ to $\{|K'_i| = 80 | 0 \le i \le n\}$ so as to reduce communication costs. We further break the framework down to two phases, *Initialization* and *Authentication* as follows.

5.4.3.1 Initialization

Foremost, aircraft planning to use ADS-B need to complete registration with TTP, where the relevant $ICAO$ is submitted to TTP by the aircraft via a secure channel established on the communication subsystem CPDLC. TTP will then handle the registration request as follows receiving $ICAO$:

- *Step1*: Taking the security level λ as input, TTP firstly acquires T and $K_F \in \{0, 1\}^{128}$ for the aircraft by calling $\Theta(\lambda)$.
- *Step2*: TTP obtains $pid \leftarrow E(ICAO, K_F, T)$.
- *Step3*: TTP randomly chooses $K_n \in \{0, 1\}^{128}$ and recursively runs the one-way function $F(\cdot)$ until obtaining the complete $Keychain = \{K_0, K_1, ..., K_n\}$.
- *Step4*: TTP sends $\sigma_1 = (pid, Keychain)$ to the aircraft through the secure channel and publicly issues $\sigma_2 = (pid, K_0)$.

5.4.3.2 Authentication

Figure 5.3 shows that we use the FFX-encrypted pid to substitute the plaintext $ICAO$; Fig. 5.4 illustrates that we concatenate M_i representing the ADS-B data to be sent with K'_{i-1}, which forms $D_i = \langle M_i || K'_{i-1} \rangle$. The keyed-hash MAC is then calculated as $\Gamma_i = H(K'_i, D_i)$.

For simplification, we reformulate the ith message as $\Omega_i = \langle Head || pid || Data_i || PC_i \rangle$ where $Head = \langle DF || CF \rangle$, $Data_i = \langle M_i || K'_{i-1} || \Gamma_i \rangle$.

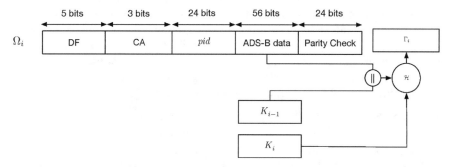

Fig. 5.4 Resistance to active attacks

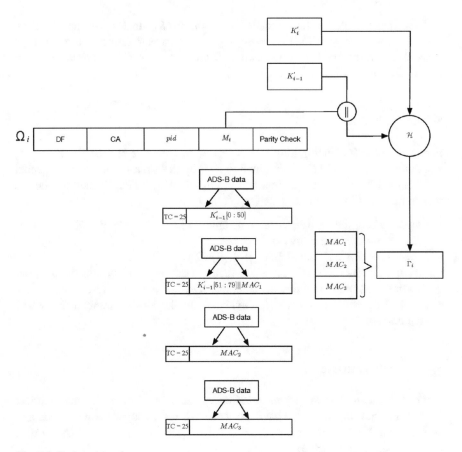

Fig. 5.5 Packet generation

As mentioned before, not including 5-bit type code, one ADS-B message remains 51-bit available space for transmitting application data, and the available space in a message cannot accommodate the too-long $Data_i$ in our scheme. Therefore, P_i is divided into 5 succeeding messages, in which the first one, except for replacing $ICAO$ with pid, is nearly the same as the original ADS-B message still filling the $Data$ field with M_i, while the $Data$ fields in the subsequent four messages are filled in sequence with $\langle K'_{i-1} || \Gamma_i \rangle$, but their type codes are both assigned to be 25 representing reserved field. Figure 5.5 details the package generation process.

- *Step1*: The broadcaster B uses pid to substitute $ICAO$.
- *Step2*: B, from $Keychain$, derives K_{i-1} and K_i, and then truncates them as $K'_{i-1} = F'(K_{i-1})$ and $K'_i = F'(K_i)$, respectively.
- *Step3*: B constructs P_i as $\langle Head || AA_i || Data_i || PC_i \rangle$ where $Data_i = \langle M_i || K'_{i-1} || \Gamma_i \rangle$ with $\Gamma_i = H(K'_i, \langle M_i || K'_{i-1} \rangle)$, and then orderly fills the

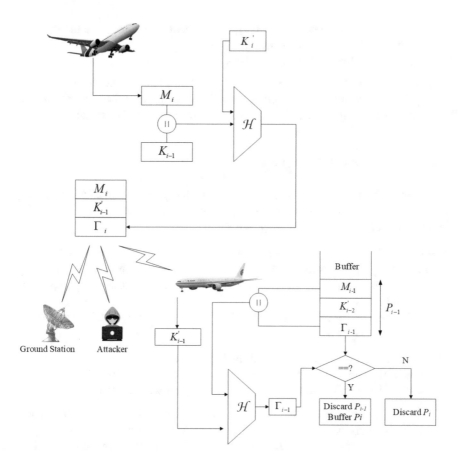

Fig. 5.6 Packet verification

Data field with $\langle K'_{i-1}||\Gamma_i\rangle$. Note that this filling occupies 4 messages all with $TC = 25$.

- *Step4*: B sends P_i through ADS-B-Out.

Subsequently, the recipient R checks whether the received P_i conforms to the specified ADS-B message format. For example, whether each field has a specified length and whether the CRC code of the entire message is valid. If yes, then R caches P_i, otherwise just rejects it. R does not immediately validate P_i due to the delayed-authentication principle of TESLA. At this time, R does not know the K'_i, which is utilized to calculate the verification code Γ_i until R receives P_{i+1} containing K'_i. Figure 5.6 describes the verification process.

- *Step1*: The receiver firstly extracts $D_i = \langle M_i||K'_{i-1}\rangle$ and K'_i from P_i and P_{i+1}, respectively, calculates $F'(F(K'_i))$, and then compares it with K'_{i-1} in D_i. If they are equal, the procedure would go to the next step. Otherwise the receiver discards simply P_{i+1} without any other processing.

- *Step2*: The receiver evaluates $\Gamma'_i = H(K'_i, D_i)$.

- *Step3*: The receiver withdraws Γ_i from the buffered P_i and checks if $\Gamma'_i \overset{?}{=} \Gamma_i$.

- *Step4*: If the above equation holds, then P_{i+1} would be pushed into the buffer after P_i being popped off, and P_{i+1} would be thrown away otherwise.

5.4.3.3   Toleration of Packet Loss

Considering that only packet P_{i+m} are received by the recipient and all other packets (P_i to P_{i+m-1}) have been lost for unknown reasons, the validation procedure of the buffered P_i consists of the following four steps.

- *Step1*: The receiver retrieves $D_i = \langle M_i || K'_{i-1} \rangle$ and K'_{i+m-1} from P_i and P_{i+m}, respectively, evaluates recursively $F'(F^m(K'_{i+m-1}))$, and compares it with K'_{i-1} in D_i. If equal, then the receiver would proceed with the following steps, and discard directly P_{i+m} otherwise.

- *Step2*: The receiver derives $K'_i = F'(F^{m-1}(K'_{i+m-1}))$ firstly, and then calculates $\Gamma'_i = H(K'_i, D_i)$.

- *Step3*: The receiver withdraws Γ_i from P_i and checks if $\Gamma'_i \overset{?}{=} \Gamma_i$.

- *Step4*: If they are equal, P_i would be firstly popped off, and then P_{i+m} would be reserved in the buffer. In the other case, the receiver would cast P_{i+m} away.

It is worth emphasizing that it is necessary to highlight some special packet losses. Particularly considering the complicated ADS-B communication environment, the last four messages in P_i and the first message in P_{i+1} are likely to be lost because of electromagnetic interference. The recipient will probably receive only the first message in P_i and the last four messages in P_{i+1} and then reconstruct a new package by combining them in sequence. Due to the time continuity of these five messages, the reconstructed packet seems to be consistent with the ADS-B message format, but P_{i-1} authentication will fail and notify the incorrect warning message. Here, the recipient will abandon the five unmatched messages and wait for the next packet to arrive.

5.5   Extended Discussion

The basic approach described so far has shown how to protect ADS-B transmissions. This section aims to improve the suitability of our solution for deployment in practical ADS-B systems. We will address the following two points.

5.5.1 Disorder

In the proposed solution, the sender broadcasts an orderly series of packets P_1, P_2, $\cdots$, and P_n. Each packet begins with the initial ADS-B packet, followed by four other packets with the reserved field $TC = 25$ which shows the transmission of the key and MAC. In normal cases, the packets can be received in the order in which they are sent. However, since environmental variables (e.g., electromagnetic interference, climate variation, and multi-path effect) may affect the propagation velocity of ADS-B signals, it is conceivable for messages to arrive out of order. This section will discuss the disorder issue in real-world aircraft scenarios. We consider the following two categories of disorders: *Intra-packet disorder* and *Inter-packet disorder*.

- *Intra-packet disorder*: The five ADS-B messages included in a single packet may be received in the incorrect sequence.
- *Inter-packet disorder*: Earlier packets may be received later.

To address the disorder issue, we show the timestamp method based on the physical ADS-B signals' arrival time. The aircraft using this technique must provide the GPS timestamp in the ADS-B transmission. To resolve intra-packet disorder, it is just a matter of restoring the order of ADS-B messages through the timestamp. In light of inter-packet disorder, the technique may seem very complex since the timestamp may determine the orientation of deriving the keys required to authenticate packets in the Keychain. In specific, for *Inter-packet disorder*, although the recipient has received P_{i-1}, P_{i+1} may arrive earlier than P_i. In this case, verifying P_{i-1} requires extracting the keys twice from K'_i in P_{i+1} to K'_{i-2} in P_{i-1} by $K'_{i-2} = F'(F^2(K'_i))$. Therefore, after using buffered P_{i+1} to authenticate P_{i-1}, due to the one-way function of $Keychain$, the recipient will no longer use P_{i+1} to verify later arrived P_i but directly discard the disordered P_i as the damaged one.

5.5.2 Adaptive-TESLA

As described before, the retroactive key in $Keychain$ is issued to authenticate ADS-B messages in a delayed manner. Specifically, considering the ever-congested 1090ES data link, each $K_i \in \{0, 1\}^{128}$ in $Keychain$ is required to be truncated into $K'_i \in \{0, 1\}^{80}$, which also lower communication costs. Nonetheless, to authenticate M_i, K'_{i-1} still needs to be transmitted along with M_i every time. In this case, if we select HMAC-MD5-96 as the keyed-hash function and transmit just a message with a payload of 51 bits, we are forced to pay an additional 176 bits, which would consume the additional bandwidth of four messages after the original one of carrying M_i.

The advantage of this onetime usage of distinct keys for each packet is that it ensures faster verification and resilience to memory-based DoS attacks since the

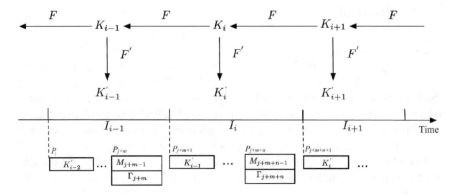

Fig. 5.7 Adaptive-TESLA

receiver is only required to buffer one packet needed to validate subsequent arrivals. The drawback is obvious since each time the key must be sent along with the message, resulting in a significantly increased communication cost proportional to the number of transmitted messages. To minimize this overhead, we devised an innovative adaptation of the TESLA, dubbed Adaptive-TESLA, that requires ADS-B participants to be in loose temporal synchronization, which may be readily accomplished through satellite-based navigation systems like as GPS. Since a result, the sender would not be required to transmit each packet accompanied by a unique key each time, as the same key would create separate MACs for consecutively emitted messages only if they occurred within the same time period, significantly reducing the traffic load.

Specifically, according to the specification of 1090ES, each of $TC = 25, 26, 27, 30$ represents the reserved field. In this case, *Adaptive-TESLA* may select $TC = 25$ and $TC = 26$ to differentiate the MAC packet from the key packet. Figure 5.7 shows that at the beginning of I_i, the sender fills K'_{i-1} in the *Data* field with $TC = 26$ indicating key delivery; subsequent packets with $TC = 25$ denoting MAC delivery, will no longer fill K'_{i-1} in I_i, thereby greatly reduce communication overheads.

5.6 Security Analysis

5.6.1 Privacy

The most simple and apparent security weakness in ADS-B communications is for an attacker to collect wireless ADS-B broadcasts via unencrypted data links using only cheap receivers. In that scenario, the adversary can eavesdrop on and collect enough clear ADS-B messages and tries to establish a correlation between the aircraft's $ICAO$ and its corresponding locations, jeopardizing the privacy of the aircraft's $ICAO$ and corresponding locations, jeopardizing the aircraft's privacy, for example, business interests or personal preferences. It is necessary to maintain the openness of ADS-B systems, ensuring that aircraft location data remains publicly accessible. As a result, our countermeasure is to use FFX to encrypt just the aircraft's unique digital $ICAO$ identification rather than the whole ADS-B transmission. As a result, such a connection is eliminated, and the identification of an aircraft cannot be associated with its exact geographic information. Thus, our approach achieves both privacy and transparency.

This also implies that our solution's privacy would be based on FFX security. As is well known, the ciphertext in FFX follows the same format as the plaintext. However, the plaintext comparable patterns have been sufficiently dispersed throughout the ciphertext to preclude potential cryptanalysis, such as known plaintext assaults. Indeed, the semantic security against adaptive chosen-ciphertext attacks has been verified in FFX, and the premise is that the underlying round function (like AES) is an advisable pseudo-random function. As a result, our approach may offer robust confidentiality safeguards while sustaining the integrity of aircraft data in the ADS-B environment.

5.6.2 Authenticity and Integrity

A skilled adversary intending to forge or changing ADS-B data carries aircraft Ghost Injection attacks out. However, it is impractical to generate proper MACs for ADS-B communication when the keys in the Keychain are adversary-agnostic. Even if an attacker gets a single key, the adversary may use it to generate MAC only once, since a one-way Keychain is not negatively derivable, thus confirming our solution's informally active-attack-resistance. Following that, we will establish the formal validity and integrity of distributed ADS-B packets based on three assumptions:

Assumption 1: There is no PPT algorithm that can distinguish between a pseudo-random function (PRF) and an ideal random function.

Assumption 2: H is secure with collision resistance if H is an algorithm of keyed-hash message code (HMAC) [19].

Assumption 3: A hash function exhibits target collision resistance (TCR); that is, given a fixed message x and a hash function $F(\cdot)$, it is found that $x' \neq x$ makes $F(x') = F(x)$ is computationally impractical [20].

Theorem 5.1 *Under the premise that a PRF is indistinguishable from an ideal random function, H is safe with collision resistance, and $F(\cdot)$ is a hash function with TCR property, the proposed technique may guarantee the authenticity and integrity of ADS-B data packets.*

Proof Here, we only give a sketchy proof framework on the basis of Theorem A.1 in [21]. First of all, we assume that H is secure with collision resistance and $F(\cdot)$ is a hash function with TCR property. In that case, the authenticity and integrity of our proposal depends mainly on indistinguishability between a PRF and a totally random function.

First, assume there exists an attacker A who can destroy the authenticity and integrity of the proposed solution, that is, A sends a m to a receiver R so that R, with non-negligible advantage, believes that the m comes from a sender S and does not alter during the transmission, even if S does not deliver it. Then there exists an attacker B who uses A to defeat the indistinguishability between the true random function and PRF with non-negligible advantage.

To this end, due to the ability of B to control the communication channel, B can provide a network simulation environment for A, and then calls A in a similar way with [21]: e.g., let n be the maximum number of transmitted packages, then B is required to randomly choose an integer $l \in \{1, \ldots, n\}$. In this case, A can violate such authenticity and integrity, that is, A, by an interaction with B, can successfully forge the lth packet P_l.

Note that the purpose of B is to differentiate truly random functions from PRFs, and B is provided with access an oracle $G(\cdot)$ in the interaction game. Therefore, after A issues a selected query m to B, B will send it to the oracle $G(\cdot)$. Finally, A gets an answer from $G(m)$, which is a pseudo-random number $PRF(M)$ or a truly random number evenly distributed in $\{0, 1\}^*$. After executing the query, B is required to determine whether $G(\cdot)$ is a real random function or just a PRF. If not incorrect, B will triumph in the game. We then made an argument that B would prevail the game if A could falsify the ADS-B package with a non-negligible advantage.

If $G(\cdot)$ is ideally random, A has little advantage in the successful fabrication of ADS-B data transmission. However, as aforementioned, the probability ϵ for A to forge P_l is not imperceptible if the packet is authenticated using PRF. Therefore, B succeeds in the game by at least ϵ / l advantage, which is still not trivial. What is more, it is also impossible for A to submit a false initial packet P_1 to R in accordance with *Assumption 2*. In addition, if A can deceive R into accepting a spoofing packet P_l, this means that a collision with $F(P_l') = F(P_l)$ may be found, which disobeys *Assumption 3*. Ultimately, we can get the conclusion from these contradictions that our proposed scheme indeed guarantees the authenticity and integrity of ADS-B packets.

5.6.3   Security Comparison

As shown in Table 5.2, we compare the security level to that of current approaches: (1) ECDSA, which is based on the elliptic curve cipher (ECC), is used to ensure the integrity of ADS-B communication by directly appending ECC signatures to the end of related messages. This may cause high costs associated with changing ADS-B protocols or updating older equipment. Meanwhile, Pan's work would continue to broadcast all fields of the ADS-B message in plain text, breaching privacy. (2)

Table 5.2 Security functions of proposed scheme

Security function	ECDSA [12]	TESLA [22]	FFX [23]	Ours
Privacy			✓	✓
Integrity	✓	✓		✓
Compatibility			✓	✓

TESLA preserves the integrity of ADS-B communication rather than their privacy, and it needs the same manner that the MAC is added to the message, thus failing to ensure compatibility. (3) Due to the format-preserving encryption features, FFX can provide both privacy and compatibility for the ADS-B system. However, it does not provide integrity protection, which may result in security vulnerabilities, such as flight data corruption or the insertion of ghost aircraft. Unlike prior approaches, ours achieves privacy, integrity, and compatibility all at the same time.

5.7 Performance Evaluation

We will assess the performance of our solution in an ADS-B scenario in two phases: Initialization, which involves the creation of $Keychain$ and pid, and Online Authentication, which includes ADS-B message verification. This evaluation would be conducted on two types of aviation participants: aircraft equipped with resource-constrained avionics that can be emulated using a smart phone equipped with an ARM-v7 processor running the Android 5.0.1 operating system, and air traffic controllers (ATCOs) equipped with powerful computers that can be emulated using a server equipped with an E5-2620@2.40 GHz processor running the Windows Server 2008 R2 operating system. Using real ADS-B data that comes from the OpenSky sensor network makes this simulation unique [24].

5.7.1 Encryption

During the Initialization phase, the ATCO encrypts the aircraft's true identity $ICAO$ using FFX to get the matching pseudonym pid, and then substitutes the clear $ICAO$ with the produced ciphertext pid. When compared to the time required for FFX encryption, the time required for replacement may be disregarded. The time required to create $Keychain$ is minimal since one-way hash operations are quick. As a result, the time overhead associated with initialization is mostly because of the FFX encryption. As a result, we would estimate the encryption time as follows, based on the number of planes

5.7.2   Authentication on ATCO

Because the number of aircraft changes considerably due to unique airspace, we would measure the encryption time on ATCO for every 1000 aircraft between 1000 and 10,000 to properly represent the time cost of FFX in real-world ADS-B systems. The experimental findings are summarized in Fig. 5.8, showing that encryption time is approximately linear with aircraft count. To illustrate the effectiveness of ADS-B encryption intuitively, we also compute the average time for one aircraft, which comes to about 30 ms. As a result, the encryption in Initialization is very efficient, and ATCO can concurrently process registrations for many participants, which is appropriate for a large-scale ADS-B network.

In our proposal, aircraft must continuously send ADS-B broadcast messages and associated MACs to nearby nodes, including ground stations linked to ATCO and other aircraft on the route. As a result, all listening receivers must validate ADS-B signals received in order to avoid potential injection and modification attacks. As a result, the performance of authenticating many messages must be assessed in order to determine the availability of our methods when applied to a large-scale air traffic network. Participants, however, are likely to possess unique processing skills. For example, although airplanes typically have resource-constrained integrated avionics, the ATCO may be outfitted with high-performance servers. As a result, this assessment must consider the processing capabilities of the receivers. This section will begin by analyzing the authentication performance of ATCO.

We simulate authentication on ATCO using the Python programming language and gather a large quantity of data on spent time, culminating in Fig. 5.10, which

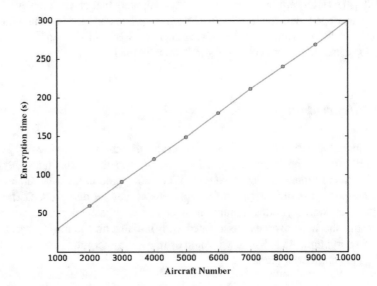

Fig. 5.8 Encryption time of ATCO

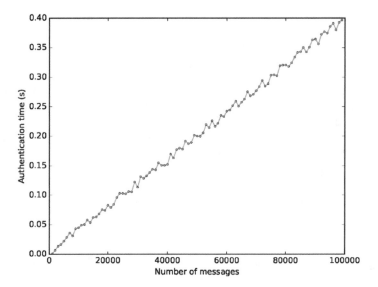

Fig. 5.9 Authentication time of ATCO

depicts the time cost in terms of buffered messages up to 100,000. As shown in Fig. 5.9, the time cost of verification approaches linearity as the number of messages increases, with a mean duration of about 0.004 ms for certifying a single message. As a result, our authentication method may provide relatively lightweight integrity safeguards by quickly identifying injected fake ADS-B signals, enabling it to scale to ever-increasing air traffic and aircraft density in most airspace.

5.7.3 Authentication on Aircraft

The aircraft are also needed to check incoming flight data to fight against the insertion of bogus messages, such as ghost airplanes on the cockpit display. For capability-constrained avionics, we conduct performance simulation to accurately represent the handling capabilities of various kinds of on-board devices.

Based on the explanation above, we rebuild the authentication software using Java and NDK C++ and then install it on the smart phone's ARM-v7 embedded CPU. By gaining root access to the Android system, we may change the processor frequencies to match our computing power levels by launching APPs, such as SETCPU. As shown in Fig. 5.10, the CPU frequency ranges from 2265 MHz to 300 MHz. The lower the CPU frequency, the lower the authentication efficiency. However, even at 300 MHz, the lowest frequency, the time required to validate ADS-B messages, is only 0.018 ms. As a result, our approach is suitably light, and the overhead of authenticating time would place no strain on either the ATCO or the aircraft, thus easing the need for computing resources compatible with a range of widely accessible avionics.

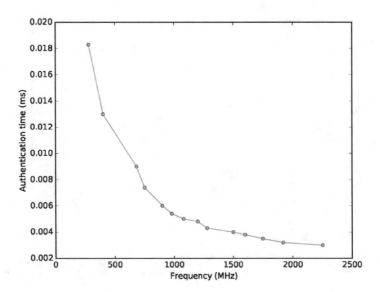

Fig. 5.10 Authentication time of aircraft

5.8   Compatibility Analysis

The suggested solution's primary aim is to ensure compatibility in a real-world air traffic monitoring scenario. As a result, we will provide the findings of our experiment on compatibility during real-world deployment in this section. To help you understand this better, we would like to begin with a brief explanation of why our approach can achieve compatibility in the following.

- *FFX*: Compared to traditional encryption algorithms, when using FFX in an ADS-B environment, the format-preserving property ensures that the aircraft's real digital identifier $ICAO$ and its pseudonym pid have the same format, including the length and character repertoire, without altering the message format. Using standard block ciphers such as AES would need additional padding to comply with the fixed block size restriction, further straining the already crowded 1090ES channel.
- *TESLA*: Our methods for ADS-B message verification are based on the TESLA standard, by filling in the K' and MAC in the reserved fields of Data with $TC = 25$ or $TC = 26$, that may comply with current ADS-B requirements. As a result, when current transponders receive our authenticating messages, they do not discard them as contaminated, but interpret them as reserved and forward them to high-level applications for further processing, avoiding the need to update old systems.

To evaluate compatibility in a practical air traffic surveillance environment, we carry out the proposed solution at the Chongqing Jiangbei International Airport, as

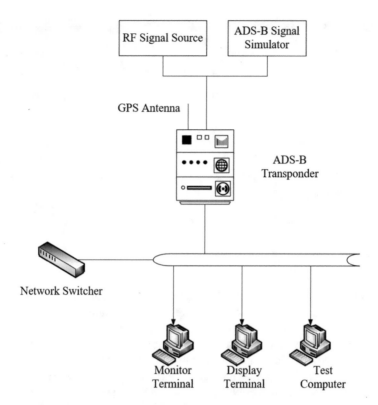

Fig. 5.11 Test environment of proposed solution

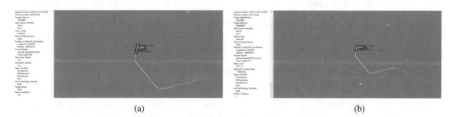

Fig. 5.12 Trajectories with/without encryption. (**a**) Trajectory with encryption. (**b**) Trajectory without encryption

shown in Fig. 5.11. Specifically, we reassemble the original ADS-B messages from the Chongqing Jiangbei International Airport based on our solution and broadcast the reassembled ones through the 1090ES data link. The ADS-B receiver parses the reassembled data according to the ADS-B message format and determines whether the parsed data conforms to the format. Noncompliance will result in the data being discarded or otherwise displayed on the monitoring screen.

The tests of our solution involve processing the encrypted message and the keyed-hash MAC on off-the-shelf hardware, as illustrated in Figs. 5.12a,b. We set

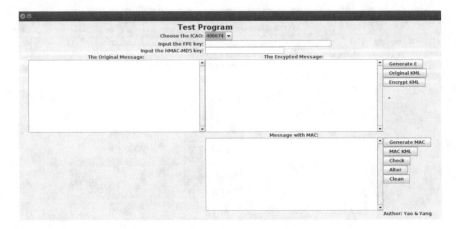

Fig. 5.13 Test program of proposed solution

the original ICAO address as "406674" and the FFX encrypts this plaintext address
to the ciphertext address "274ba7". These two ICAO addresses have the same
format, and the transponder cannot effectively distinguish between them. We use
the same air traffic data to calculate two trajectories, with or without MAC. These
two trajectories look the same on the monitoring screen, which proves that the
current ADS-B equipment does not discard our reassembled ADS-B messages. As
a result, our solution is compatible with current ADS-B protocols and transponders.
Figure 5.13 shows the graphical user interface (GUI) of our test program. As shown
in Fig. 5.14, we list the running steps and results of our test program as follows:

- *Step1:* Click the "Clean" button to delete the data generated before.
- *Step2:* Input the FPE key like "123457890abcdef1234567890abcdef" and the
  MAC-MD5 key like "1234567890abcdef1235". Notice that the characters contained in these keys should be in the format of hex.
- *Step3:* Click the "Generate E" button to encrypt the original ADS-B message,
  where the original ICAO address is set to "406674" and the resulting *pid* is
  assigned to "274ba7" after FFX encryption. Nevertheless, we need to recalculate
  the parity-check code accordingly.
- *Step4:* Click the "Generate MAC" button to create the MAC of the original ADS-
  B message, and the resulting MAC is sequentially encapsulated in four ADS-B
  messages.

It is worth mentioning that if you have Google Earth on your computer, you can
click those "KML" related buttons to check the trajectories in Google Earth.

To assess our proposal's effectiveness in a real-world ADS-B environment, we
conduct authentication tests at the aforementioned airport, which is very busy,
particularly during rush hour, with about 200 aircraft taking off and landing. The
result shows that when the transponder operates at a frequency of 300 MHz, the
total time required for authentication is just 81 ms for up to 4500 messages. As a

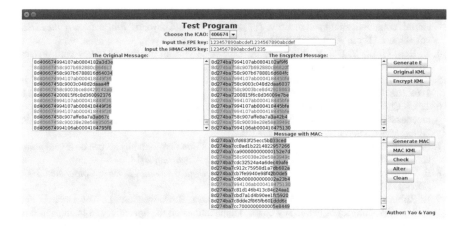

Fig. 5.14 Results of test program

result, under actual air traffic circumstances characterized by high aircraft density, our approach to ADS-B security will not degrade security or performance when the 1090ES data connection is being used, making it ideal for large-scale and low-cost deployments.

5.9   Conclusion

In this chapter, we have provided a complete encryption solution to ensure the privacy and integrity of ADS-B communication concerning security requirements in real-world ADS-B environments. Simultaneously, the compatibility of our solution with existing ADS-B systems can be ensured through the use of FFX encryption and reserved ADS-B fields. Extensive tests using real-world flight data show the excellent performance of our solution, making it deployable in the practical ADS-B system. Other security issues, such as secure location estimation using a non-cryptographic method, will be further investigated in future work.

References

1. M. Strohmeier, V. Lenders, and I. Martinovic, "On the security of the automatic dependent surveillance-broadcast protocol," *IEEE Communications Surveys & Tutorials*, vol. 17, no. 2, pp. 1066–1087, 2015.
2. C. Finke, J. Butts, and R. Mills, "ADS-B encryption: confidentiality in the friendly skies," in *Proceedings of the Eighth Annual Cyber Security and Information Intelligence Research Workshop*. ACM, 2013, pp. 1–9.

3. K. Sampigethaya, R. Poovendran, S. Shetty, T. Davis, and C. Royalty, "Future e-enabled aircraft communications and security: The next 20 years and beyond," *Proceedings of the IEEE*, vol. 99, no. 11, pp. 2040–2055, 2011.
4. H. Ren, H. Li, Y. Dai, K. Yang, and X. Lin, "Querying in internet of things with privacy preserving: Challenges, solutions and opportunities," *IEEE Network*, pp. 1–8, 2018.
5. K. Sampigethaya and R. Poovendran, "Privacy of future air traffic management broadcasts," in *2009 IEEE/AIAA 28th Digital Avionics Systems Conference*, Oct 2009, pp. 6.A.1–1–6.A.1–11.
6. T. C. Chen, "An authenticated encryption scheme for automatic dependent surveillance-broadcast data link," in *Proceedings of Cross Strait Quad-Regional Radio Science and Wireless Technology Conference*, 2012, pp. 127–131.
7. K. D. Wesson, T. E. Humphreys, and B. L. Evans, "Can cryptography secure next generation air traffic surveillance?" *IEEE Security and Privacy Magazine*, 2014.
8. E. Valovage, "Enhanced ADS-B research," in *25th Digital Avionics Systems Conference, 2006 IEEE/AIAA*. IEEE, 2006, pp. 1–7.
9. R. Robinson, M. Li, S. Lintelman, K. Sampigethaya, R. Poovendran, D. Von Oheimb, and J.-U. Buber, "Impact of public key enabled applications on the operation and maintenance of commercial airplanes," in *7th AIAA ATIO Conf, 2nd CEIAT Int'l Conf on Innov and Integr in Aero Sciences, 17th LTA Systems Tech Conf; followed by 2nd TEOS Forum*, 2007, p. 7769.
10. M. J. Viggiano, E. M. Valovage, K. B. Samuelson, and D. L. Hall, "Secure ADS-B authentication system and method," Jun. 1 2010, US Patent 7,730,307.
11. Z. Feng, W. Pan, and Y. Wang, "A data authentication solution of ADS-B system based on X. 509 certificate," in *27th International Congress of the Aeronautical Sciences, ICAS*, 2010, pp. 1–6.
12. W. Pan, Z. Feng, and Y. Wang, "ADS-B data authentication based on ECC and X. 509 certificate," *Journal of Electronic Science and Technology*, vol. 10, no. 1, pp. 51–55, 2012.
13. A. Shamir, *Identity-Based Cryptosystems and Signature Schemes*. Springer Berlin Heidelberg, 1984.
14. H. Yang, H. Kim, H. Li, E. Yoon, X. Wang, and X. Ding, "An efficient broadcast authentication scheme with batch verification for ADS-B messages," *KSII Transactions on Internet and Information Systems*, vol. 7, no. 10, pp. 2544–2560, 2013.
15. H. Yang, R. Huang, X. Wang, J. Deng, and R. Chen, "EBAA: An efficient broadcast authentication scheme for ADS-B communication based on IBS-MR," *Chinese Journal of Aeronautics*, vol. 27, no. 3, pp. 688–696, 2014.
16. S. Zhang, H. Li, Y. Dai, J. Li, M. He, and R. Lu, "Verifiable outsourcing computation for matrix multiplication with improved efficiency and applicability," *IEEE Internet of Things Journal*, DOI: https://doi.org/10.1109/JIOT.2018.2867113, 2018.
17. H. Li, Y. Yang, Y. Dai, J. Bai, S. Yu, and Y. Xiang, "Achieving secure and efficient dynamic searchable symmetric encryption over medical cloud data," *IEEE Transactions on Cloud Computing*, vol. PP, no. 99, pp. 1–1, 2017.
18. M. Schäfer, V. Lenders, and I. Martinovic, "Experimental analysis of attacks on next generation air traffic communication," in *International Conference on Applied Cryptography and Network Security*. Springer, 2013, pp. 253–271.
19. M. Bellare, "New proofs for NMAC and HMAC: security without collision-resistance," in *Proceedings of International Conference on Advances in Cryptology*. Springer, 2006, pp. 602–619.
20. M. Bellare and P. Rogaway, "Collision-resistant hashing: Towards making UOWHFs practical," in *Proceedings of Advances in Cryptology — CRYPTO '97*. Springer, 1997, pp. 470–484.
21. A. Perrig, J. D. Tygar, D. Song, and C. Ran, "Efficient authentication and signing of multicast streams over lossy channels," in *Proceedings of IEEE Security and Privacy*, 2000, pp. 56–73.
22. A. Perrig, D. Song, R. Canetti, J. Tygar, and B. Briscoe, "Timed efficient stream loss-tolerant authentication (tesla): Multicast source authentication transform introduction," Tech. Rep., 2005.

23. M. Bellare, P. Rogaway, and T. Spies, "The FFX mode of operation for format-preserving encryption," *NIST submission*, vol. 20, 2010.
24. M. Strohmeier, I. Martinovic, M. Fuchs, M. Schäfer, and V. Lenders, "Opensky: A Swiss army knife for air traffic security research," in *Digital Avionics Systems Conference (DASC), 2015 IEEE/AIAA 34th*. IEEE, 2015, pp. 401–413.

Chapter 6
Conclusion and Future Work

6.1 Conclusion

From 2020, all aircraft must be equipped with ADS-B transponders when entering European and American airspace [1, 2]. Although ADS-B has replaced radar as the core of next-generation ATC, ATC still uses radars to cross-validate surveillance data. We can compare ADS-B signals with radar signals to make sure that flight data is correct, so combining ADS-B and radar surveillance will enhance flight safety. Nevertheless, to give full play to the advantages of ADS-B, it is still necessary to reduce the dependence on radar monitoring. To improve air traffic safety further, we also need to take more measures to prevent security threats to ADS-B communications since ADS-B messages are transmitted on public broadcast channels and are not encrypted or authenticated. Attackers may intercept ADS-B messages or forge ghost aircraft using the low-cost SDR. To minimize the harm caused by these attacks, researchers have suggested many methods, such as authentication, encryption, and location verification [3–5]. In this monograph, we propose efficient and compatible ADS-B broadcast authentication, message privacy protection, and aircraft location verification schemes, which fully ensure ADS-B security and are suitable for large-scale practical deployment. Specifically, based on the analysis and discussion provided throughout this monograph, we present the following remarks:

– Analyzing ADS-B vulnerabilities can help readers fully understand ADS-B security issues. Aviation experts, computer security analysts, and government officials can then better realize these vulnerabilities and come up with ideas for viable ADS-B security solutions. This also contributes to exposing the flaws of NextGen and setting off a wave of advocacy for vigorously protecting ADS-B. In particular, we show how to maintain the security of ADS-B so that we can use this as a guide for the aviation sector and government agencies to make air traffic safety decisions.

© The Author(s), under exclusive license to Springer Nature Switzerland AG 2023   143
H. Yang et al., *Secure Automatic Dependent Surveillance-Broadcast Systems*,
Wireless Networks, https://doi.org/10.1007/978-3-031-07021-1_6

– The security technology used in ADS-B needs to meet two constraints: (1) the ADS-B system must remain open, and (2) changes to ADS-B must be economically feasible. However, existing technologies cannot meet these limitations. First, although adding authentication to the existing ADS-B system does not require modification of the physical layer, it cannot solve the problems of interference and deception. Second, the encryption used by ADS-B has two major flaws: one is that it contradicts the openness of ADS-B, and the other is that there are many participants in air traffic and the cost of key management is high. Third, location verification methods (such as MLAT) have high success rates but are expensive to deploy.
– Modern cryptography has proven to be a mature technology for securing wireless communications. ADS-B wireless broadcasts exhibit some distinct characteristics from traditional wireless communications, such as huge distances, outdoor line-of-sight, and few multi-path effects. This naturally raises the question: *Can cryptography protect ADS-B?* In this monograph, we examine this question and propose several effective cryptographic solutions. To evaluate the effectiveness of our solutions, we carry them out at Chongqing Jiangbei International Airport. Tests at this airport have shown significant security improvements for ADS-B communications.

6.2 Future Work

Although cryptography-based methods can effectively enhance the security of ADS-B, the implementation of cryptography enhancements may also face significant regulatory and technical complexity. Regulators and commercial airlines may not be able to accept the burden of public-key management and the decline in operating capacity for 1090ES. To avoid these limitations, a possible alternative is to use non-cryptographic techniques. Below, we suggest some future research directions that will utilize non-cryptographic techniques to guarantee ADS-B security.

6.2.1 Secure Trajectory Validation and Prediction

6.2.1.1 Trajectory Validation

As discussed in Chap. 4, the air location verification techniques can validate claimed locations, allowing for the detection of attacks such as the insertion of fake location data. These techniques, however, entail significant system requirements including tight time synchronization, specialized hardware, or the modification of existing protocols. Additionally, the location verification technique is not designed for the aircraft's intrinsic agility. Considering the slow adoption of new technologies in the aviation industry, it is reasonable to conclude that proactive solutions will not be used to solve ADS-B security problems for a long period thereafter. Rather than that, passive solutions are needed that do not involve a change to the existing infrastructure.

The secure trajectory validation technique is a potential option to meet these requirements, which is in line with the ADS-B standard [6]. In particular, it uses the mobility of the aircraft to evaluate a sequence of claimed positions rather than a single position. Because of the continuous movement of the aircraft, the distance between the aircraft and the receiver is constantly changing, resulting in different propagation delays. The work in [7] proved that even with a single low-cost receiver, these false aircraft trajectories can be detected. It further shows that fixed attackers cannot forge these propagation delay differences for over three receivers at the same time. If an attacker broadcasts a false positional claim, considering 3 verifiers at different positions, there is no position spoofed by the attacker that cannot be detected. Therefore, at least one receiver will identify trajectory forgery by comparing the measured differences between the actual propagation delay and the expected propagation delay for at least four different receivers. The simulation results show that, in this way, a group of trusted receivers can completely distinguish between false and true positional claims with no false positives or false negatives. In addition, each receiver can measure independently, so there is no need for time synchronization between receivers or transmitters. The simplicity of the solution makes it particularly suitable for use on aircraft, without upgrading the existing infrastructure and with only low-cost equipment. Besides, the technology has considerable flexibility. For example, a trusted aircraft equipped with ADS-B-In can also be used as a receiver for trajectory verification to extend the range and improve safety. Based on preliminary findings, mobile verifiers (such as trusted aircraft) may help to further reduce system limitations and speed up the identification of false trajectories.

6.2.1.2 Trajectory Prediction

As a core technology of NextGen, aircraft trajectory prediction (ATP) traditionally uses deterministic modeling, in which the model quality is affected by the uncertainties during different phases of the flight [8]. The expression of such uncertainties commonly leverages a probability density function (PDF), such that the input uncertainties are propagated through deterministic models to the output uncertainty. Because of the difficulties of capturing the uncertainties, these traditional techniques can only predict a single trajectory, lowering model performance [9]. With the ever-increasing aviation data, ATP uses data-driven modeling to get trajectory features from ADS-B and/or SSR data. Various machine learning algorithms are exploited to perform ATP by extracting relevant trajectory features from aviation big data [10–12].

By doing so, we can also identify the primary sources of uncertainty in ATP. The uncertainty may occur from inaccuracies in location and velocity measurement or rapidly changing weather circumstances [9]. It may also be attributed to a lack of knowledge about the operating strategy of the airline or the divergence of ATC operations [13]. Another big source of uncertainty is in the time dimension, for example, because the take-off time forecast is inaccurate [14]. Current research focuses on the evaluation of prediction accuracy, which could be described by the spatial-temporal errors between the predicted and actual trajectories [15]. Besides, the quantification of prediction uncertainties remains a problematic topic for ATP. In [16], the authors offer a two-stage Gaussian process regression (GPR) technique for assessing trajectory prediction errors and uncertainty. One GPR model analyzes historical flights with comparable flight patterns, while another incorporates the partially flown trajectories of a flight. Finally, the GPR models produce prediction distributions for quantifying the uncertainty.

6.2.2   Authentication on Physical Layer

Non-cryptographic techniques such as fingerprinting cover various identity authentication and device identification methods on the physical layer [17–20]. Its goal is to determine whether suspicious behavior has occurred on the ADS-B communication. If there is a difference between legal and illegal data packets on the physical layer, machine learning algorithms can build models to predict suspicious behaviors. Even if only the device type can be identified but the participating user cannot be identified, this also helps to identify the intruder. Specifically, fingerprinting can be used for ADS-B broadcast authentication on the physical layer based on the unique characteristics of hardware or software [21]. In addition, ADS-B signals are transmitted on the wireless broadcast channel, which can conduct authentication based on the unique characteristics of channels. As a result, fingerprinting may be divided into three categories: software-based, hardware-based, and channel-based [21].

- *Software-based fingerprinting*. This technique takes advantage of the behavioral distinctiveness of software running on different avionics. If the firmware and drivers of different avionics are sufficiently distinctive, this kind of fingerprinting is workable. However, given that major aircraft manufacturers use the same firmware and drivers in their production processes, the differences in software fingerprints are negligible. Unlike Wi-Fi, which has many suppliers, the commercial aircraft market is dominated by two large companies (Boeing and Airbus). In the long run, the substantial difference between the software of ADS-B suppliers is not significant. Therefore, it is usually unlikely that this method will be used for authentication and identification of ADS-B devices.
- *Hardware-based fingerprinting*. This technique identifies wireless devices by detecting a small difference in the hardware. It creates a unique fingerprint for specific hardware by comparing on/off transients or modulation of RF signals. This technique can capture signals within a range of 15 m. But based on replay messages, an attacker can use SDR to bypass the effectiveness of this

technique. Note that RF fingerprinting is a typical approach to hardware-based fingerprinting that can identify the wireless device, such as aircraft categorization [22] and VANET communication [23]. Because of manufacturing defects, the wireless components of these devices have slightly different RF features that can be treated as unique and inherent RF fingerprints [24]. Therefore, we can extract RF fingerprints to determine the identities of wireless devices [25].

In addition, this technique can also use the time difference between the clocks of the two devices (also called clock skew) to establish a unique fingerprint. To use clock skew, ADS-B communication must include a timestamp, but there is no timestamp in the FAA's pre-defined ADS-B message structure. A capable attacker might examine a single message stream and try to replicate the clock skew.

Finally, by adding additional circuitry (called a physical unclonable function) to the ADS-B transmitter, a unique and secure hardware fingerprint can be created. Although theoretically feasible, this solution requires new ADS-B hardware and switches from the broadcast mode to the challenge-and-response mode. Therefore, it is not practical to use this technique to complete hardware fingerprint-based authentication.

– *Channel-based fingerprinting.* This technique takes advantage of the inherent characteristics of the communication medium. It may be based on received signal strength, channel impulse response, carrier phase, *etc*. This technique can replace traditional authentication and verification techniques for ADS-B broadcast messages.

6.2.3 Event Detection

OpenSky can be used to identify abnormal events that occur within the coverage of the sensor network [26]. By combining the sensor data from OpenSky with the publicly available 24-bit ICAO identifier, aircraft type, and airline data, various types of activities can be tracked. Although this data cannot be directly derived from the aircraft's supplier and airline, it can be used for free on the Internet, e.g., the *Planeplotter* APP.[1] ICAO also provides the current airline of the corresponding aircraft, giving another classification feature. Here, we illustrate an example of the impact of COVID-19 on the global aviation industry [27].

OpenSky has released the flight metadata set collected throughout 2019 and 2020 for comparison considering the COVID-19 pandemic.[2] Xavier Olive used this data set to draw some initial graphs that illustrate the decline in air traffic at specific airports in the early stages of the pandemic.[3]

---

[1] http://www.coaa.co.uk/planeplotter.htm.

[2] https://doi.org/10.5281/zenodo.3737101.

[3] https://traffic-viz.github.io/scenarios/covid19.html.

Using time series analysis can provide potential insights into the impact of COVID-19 on global aviation. For example, the COVID-19 pandemic has a serious impact on aviation around the world. This influence can be reflected in the data and has some regional characteristics. In particular, we have the following observations:

- *The evolution of flights at each airport.* The current trends in the number of aircraft departing from airports in various regions around the world are: Asian airports declined slowly in February, European airports plummeted in early March, US airports fell late, and India almost stopped all air traffic.
- *The evolution of flights at each airline.* We observe that the declining pattern of scheduled airlines depends on geographic location, with more low-cost airlines stopping their activities and cargo airlines continuing.

6.2.4 Message Anomaly Detection

The goal of message anomaly detection is to identify message transmissions that do not conform to the expected pattern. In other words, anomaly detection is the identification of abnormal values, noise, and deviations. Existing anomaly detection methods are mainly divided into distance-based [28], ensemble-based [29, 30], statistics-based [31], domain-based [32], and reconstruction-based methods [33, 34]. Here, due to the specific requirements of the ADS-B scenario, the anomaly detection of ADS-B messages usually does not need to cover all these methods but mainly focuses on distance-based and reconstruction-based methods.

6.2.4.1 Distance-Based Method

This method relies on the definition of the distance/similarity function between two data instances, which is not always obvious when the data instance is not a point but more complex data (such as a time series) [28]. The distance-based method can be further divided into nearest neighbor-based [35] and cluster-based methods [36]. The former detects abnormal data points based on the distance from the abnormal data point to the neighboring point or its relative data density. The latter is an unsupervised/semi-supervised technique, which groups similar data instances into clusters according to the definition of the pairwise distance or similarity function.

Recently, a clustering method has been proposed to detect anomalies in ADS-B messages. Corrado et al. [37] firstly defined the concepts of spatial anomaly and energy anomaly, and further studied the relationship and interdependence between these two anomalies. The space measurement here refers to the latitude, longitude, altitude, and heading information of the aircraft, and the energy index is obtained from the position and speed information of the aircraft, such as potential energy, specific kinetic energy, specific energy, and speed. Moreover, spatial anomalies represent a set of air traffic flow trajectories that do not comply with standard space operations, and energy anomalies indicate air traffic flow trajectories that do not meet standard energy operations. Specifically, the HDBSCAN [38] algorithm is used to detect spatial anomalies, and the DBSCAN [39] algorithm is used to detect energy anomalies. The relationship between space anomalies and energy anomalies is obtained through experiments: under normal circumstances, if a space anomaly is detected in the aircraft trajectory, it is likely that an energy anomaly will also be detected. This conclusion can help pilots and ATCOs to choose appropriate operations when dealing with abnormal situations and avoiding potential hazards.

6.2.4.2 Reconstruction-Based Method

This method assumes that the anomaly will lose information when projected into a low-dimensional space [34]. In the case of ADS-B, this method mainly uses two neural network techniques to reconstruct the coding message: Autoencoder [40] and variational autoencoder (VAE) [41, 42]. The former has the same number of input and output neurons, and one or more hidden layers, while the latter has a smaller number of neurons to act as a compression or dimensionality reduction functionality. For anomaly detection based on reconstruction, anomalies cannot be well reconstructed from the low-dimensional representation of latent variables.

Specifically, considering that ADS-B messages are a time series, recent studies tend to use long short-term memory (LSTM) technology to detect anomalies in ADS-B messages [33, 43–46]. For example, Habler et al. [47] chose the LSTM encoder-decoder, as shown in Fig. 6.1, to analyze the information and behavior of the aircraft on the route. Specifically, the authors reconstructed the information by training the LSTM encoder-decoder model and comparing the difference between the reconstructed information and the original information to identify anomalies. The malicious information will be significantly different from the original information after reconstruction. They can then determine the type of attack by using interpretable techniques.

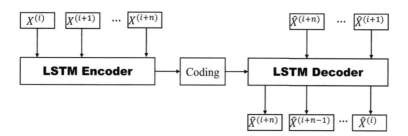

Fig. 6.1 LSTM encoder-decoder

References

1. EUROCONTROL, "Cascade news 9 - update on developments," October 2010.
2. FAA, "Automatic dependent surveillance–broadcast," October 2016.
3. H. Yang, Q. Zhou, M. Yao, R. Lu, H. Li, and X. Zhang, "A practical and compatible cryptographic solution to ads-b security," *IEEE Internet of Things Journal*, vol. 6, no. 2, pp. 3322–3334, 2018.
4. T. Kacem, D. Wijesekera, P. Costa, and A. B. Barreto, "Secure ADS-B framework "ADS-Bsec"," in *2016 IEEE 19th International Conference on Intelligent Transportation Systems (ITSC)*. IEEE, 2016, pp. 2681–2686.
5. P. Berthier, J. M. Fernandez, and J.-M. Robert, "Sat: Security in the air using tesla," in *2017 IEEE/AIAA 36th digital avionics systems conference (DASC)*. IEEE, 2017, pp. 1–10.
6. J. Sun, J. Ellerbroek, and J. M. Hoekstra, "Reconstructing aircraft turn manoeuvres for trajectory analyses using ads-b data," in *Proceedings of the 9th SESAR Innovation Days*, Dec 2019. [Online]. Available: https://www.sesarju.eu/sites/default/files/documents/sid/2019/papers/SIDs_2019_paper_57.pdf
7. M. Schäfer, V. Lenders, and J. Schmitt, "Secure track verification," in *2015 IEEE Symposium on Security and Privacy*. IEEE, 2015, pp. 199–213.
8. Á. Rodríguez-Sanz, D. Á. Álvarez, F. G. Comendador, R. A. Valdés, J. Pérez-Castán, and M. N. Godoy, "Air traffic management based on 4D trajectories: a reliability analysis using multi-state systems theory," *Transportation research procedia*, vol. 33, pp. 355–362, 2018.
9. E. Casado, C. Goodchild, and M. Vilaplana, "Identification and initial characterization of sources of uncertainty affecting the performance of future trajectory management automation systems," in *Proceedings of the 2nd International Conference on Application and Theory of Automation in Command and Control Systems*, 2012, pp. 170–175.
10. Y. Pang, N. Xu, and Y. Liu, "Aircraft trajectory prediction using LSTM neural network with embedded convolutional layer," in *Proceedings of the Annual Conference of the PHM Society*, vol. 11. PHM Society Scottsdale, AZ, USA, 2019.
11. K. Yang, M. Bi, Y. Liu, and Y. Zhang, "LSTM-based deep learning model for civil aircraft position and attitude prediction approach," in *2019 Chinese Control Conference (CCC)*. IEEE, 2019, pp. 8689–8694.
12. H. Naessens, T. Philip, M. Piatek, K. Schippers, and R. Parys, "Predicting flight routes with a deep neural network in the operational air traffic flow and capacity management system," *EUROCONTROL Maastricht Upper Area Control Centre, Maastricht Airport, The Netherlands, Tech. Rep*, 2017.
13. E. Casado, M. La Civita, M. Vilaplana, and E. W. McGookin, "Quantification of aircraft trajectory prediction uncertainty using polynomial chaos expansions," in *2017 IEEE/AIAA 36th Digital Avionics Systems Conference (DASC)*. IEEE, 2017, pp. 1–11.
14. S. Badrinath, H. Balakrishnan, E. Joback, and T. G. Reynolds, "Impact of off-block time uncertainty on the control of airport surface operations," *Transportation Science*, vol. 54, no. 4, pp. 920–943, 2020.
15. J. Bronsvoort, G. McDonald, M. Paglione, C. Garcia-Avello, I. Bayraktutar, and C. M. Young, "Impact of missing longitudinal aircraft intent on descent trajectory prediction," in *2011 IEEE/AIAA 30th Digital Avionics Systems Conference*. IEEE, 2011, pp. 3E2–1.
16. R. Graas, J. Sun, and J. Hoekstra, "Quantifying accuracy and uncertainty in data-driven flight trajectory predictions with gaussian process regression," in *11th SESAR Innovation Days*, 2021.
17. J. Hall, M. Barbeau, and E. Kranakis, "Radio frequency fingerprinting for intrusion detection in wireless networks," *IEEE Transactions on Defendable and Secure Computing*, vol. 12, pp. 1–35, 2005.
18. B. Danev, H. Luecken, S. Capkun, and K. El Defrawy, "Attacks on physical-layer identification," in *Proceedings of the third ACM conference on Wireless network security*, 2010, pp. 89–98.

19. B. Danev, D. Zanetti, and S. Capkun, "On physical-layer identification of wireless devices," *ACM Computing Surveys (CSUR)*, vol. 45, no. 1, pp. 1–29, 2012.

20. M. Leonardi and F. Gerardi, "Aircraft mode S transponder fingerprinting for intrusion detection," *Aerospace*, vol. 7, no. 3, p. 30, 2020.

21. M. Strohmeier, V. Lenders, and I. Martinovic, "On the security of the automatic dependent surveillance-broadcast protocol," *IEEE Communications Surveys & Tutorials*, vol. 17, no. 2, pp. 1066–1087, 2014.

22. M. Strohmeier, M. Smith, V. Lenders, and I. Martinovic, "Classi-Fly: Inferring aircraft categories from open data," *ACM Transactions on Intelligent Systems and Technology (TIST)*, vol. 12, no. 6, pp. 1–23, 2021.

23. M. A. Al-Shareeda, M. Anbar, S. Manickam, and A. A. Yassin, "Vppcs: Vanet-based privacy-preserving communication scheme," *IEEE Access*, vol. 8, pp. 150 914–150 928, 2020.

24. Y. Chen, H. Wen, H. Song, S. Chen, F. Xie, Q. Yang, and L. Hu, "Lightweight one-time password authentication scheme based on radio-frequency fingerprinting," *IET Communications*, vol. 12, no. 12, pp. 1477–1484, 2018.

25. L. Peng, J. Zhang, M. Liu, and A. Hu, "Deep learning based rf fingerprint identification using differential constellation trace figure," *IEEE Transactions on Vehicular Technology*, vol. 69, no. 1, pp. 1091–1095, 2019.

26. M. Strohmeier, I. Martinovic, M. Fuchs, M. Schäfer, and V. Lenders, "OpenSky: A Swiss army knife for air traffic security research," in *2015 IEEE/AIAA 34th Digital Avionics Systems Conference (DASC)*. IEEE, 2015, pp. 4A1–1.

27. R. Koelle and F. L. C. Barbosa, "Assessing the global covid-19 impact on air transport with open data," in *2021 IEEE/AIAA 40th Digital Avionics Systems Conference (DASC)*. IEEE, October 2021.

28. E. Burnaev and V. Ishimtsev, "Conformalized density-and distance-based anomaly detection in time-series data," *arXiv preprint arXiv:1608.04585*, 2016.

29. D.-I. Curiac and C. Volosencu, "Ensemble based sensing anomaly detection in wireless sensor networks," *Expert Systems with Applications*, vol. 39, no. 10, pp. 9087–9096, 2012.

30. M. Zhou, Y. Wang, A. K. Srivastava, Y. Wu, and P. Banerjee, "Ensemble-based algorithm for synchrophasor data anomaly detection," *IEEE Transactions on Smart Grid*, vol. 10, no. 3, pp. 2979–2988, 2018.

31. X. Zhu, *Anomaly detection through statistics-based machine learning for computer networks*. The University of Arizona, 2006.

32. D. Jiang, Z. Xu, P. Zhang, and T. Zhu, "A transform domain-based anomaly detection approach to network-wide traffic," *Journal of Network and Computer Applications*, vol. 40, pp. 292–306, 2014.

33. L. Basora, X. Olive, and T. Dubot, "Recent advances in anomaly detection methods applied to aviation," *Aerospace*, vol. 6, no. 11, p. 117, 2019.

34. A. Tong, G. Wolf, and S. Krishnaswamyt, "Fixing bias in reconstruction-based anomaly detection with Lipschitz discriminators," in *2020 IEEE 30th International Workshop on Machine Learning for Signal Processing (MLSP)*. IEEE, 2020, pp. 1–6.

35. X. Gu, L. Akoglu, and A. Rinaldo, "Statistical analysis of nearest neighbor methods for anomaly detection," *arXiv preprint arXiv:1907.03813*, 2019.

36. I. Syarif, A. Prugel-Bennett, and G. Wills, "Unsupervised clustering approach for network anomaly detection," in *International conference on networked digital technologies*. Springer, 2012, pp. 135–145.

37. S. J. Corrado, T. G. Puranik, O. P. Fischer, and D. N. Mavris, "A clustering-based quantitative analysis of the interdependent relationship between spatial and energy anomalies in ads-b trajectory data," *Transportation Research Part C: Emerging Technologies*, vol. 131, p. 103331, 2021. [Online]. Available: https://www.sciencedirect.com/science/article/pii/S0968090X21003351

38. L. McInnes, J. Healy, and S. Astels, "HDBSCAN: Hierarchical density based clustering," *Journal of Open Source Software*, vol. 2, no. 11, p. 205, 2017.

39. M. Ester, H.-P. Kriegel, J. Sander, X. Xu *et al.*, "A density-based algorithm for discovering clusters in large spatial databases with noise." in *kdd*, vol. 96, no. 34, 1996, pp. 226–231.
40. W. Wang, Y. Huang, Y. Wang, and L. Wang, "Generalized autoencoder: A neural network framework for dimensionality reduction," in *Proceedings of the IEEE conference on computer vision and pattern recognition workshops*, 2014, pp. 490–497.
41. D. P. Kingma and M. Welling, "Auto-encoding variational bayes," *arXiv preprint arXiv:1312.6114*, 2013.
42. J. An and S. Cho, "Variational autoencoder based anomaly detection using reconstruction probability," *Special Lecture on IE*, vol. 2, no. 1, pp. 1–18, 2015.
43. S. Hochreiter and J. Schmidhuber, "Long short-term memory," *Neural computation*, vol. 9, no. 8, pp. 1735–1780, 1997.
44. S. Akerman, E. Habler, and A. Shabtai, "VizADS-B: Analyzing sequences of ADS-B images using explainable convolutional LSTM encoder-decoder to detect cyber attacks," *arXiv preprint arXiv:1906.07921*, 2019.
45. J. Wang, Y. Zou, and J. Ding, "ADS-B spoofing attack detection method based on LSTM," *EURASIP Journal on Wireless Communications and Networking*, vol. 2020, no. 1, pp. 1–12, 2020.
46. Y. Chen, J. Sun, Y. Lin, G. Gui, and H. Sari, "Hybrid N-inception-LSTM-based aircraft coordinate prediction method for secure air traffic," *IEEE Transactions on Intelligent Transportation Systems*, 2021.
47. E. Habler and A. Shabtai, "Using LSTM encoder-decoder algorithm for detecting anomalous ADS-B messages," *Computers & Security*, vol. 78, pp. 155–173, 2018.

Index